Encyclopaedia of

Nanotechnology

Encyclopaedia of
Nanotechnology

Vol. 3 : Properties of Nanomaterials

Dr. Rita Khare

AGROTECH PRESS

JAIPUR ■ NEW DELHI

AGROTECH PRESS

Regd. Off. "Bony Residency", Gate No. 2
Opp. Tilak Public School, Vishweswariya Nagar,
Gopalpura Road, Jaipur - 302018, India
Ph: 0141 - 2761280, Telefax: 2761381
E-mail: agrotechpress@gmail.com

Delhi Office: 102, 1st Floor, Satyam House,
4327/3, Ansari Road, Darya Ganj,
New Delhi - 110002, India
Ph: 011-45652440
visit us: www.agrotechpress.com

Distributed by :
KEYA INTERNATIONAL PRESS LLP
E-mail: keyainternationalpressllp@gmail.com

Published in India, 2024

ISBN : 9789384568566 (10 Vol. Set)

Typeset by Agrotech Press

Printed and bound in India by Replika Press Pvt. Ltd.

Contents

Preface

Nanotechnology is the collaboration of the physics, chemistry, biology, computer and material sciences integrated with engineering entering the nanoscale. This means science and engineering focused on making the particles, things and devices at the atomic and molecular scale. Nanomaterials or the Nanoparticles are the set of particles or the substances where atlas one dimension is less than approximately 100nm. Nanomaterial research is a field that takes a materials science-based approach on nanotechnology. Sometimes nanomaterials are used as thin films or surface coatings, as on computer chips or as nanowires, nanotubes, or as blobs of tiny nanocrystalline particles. It studies materials with morphological features on the nanoscale, and especially those that have special properties stemming from their nanoscale dimensions. Nanomaterials and Nanotechnologies attract tremendous attention in recent researches. Nanomaterials attract tremendous attention in recent researches. The variety of nanomaterials is great, and their range of properties. Although extensive research has been done in this field it still lacks a comprehensive reference work that presents data on properties of different Nanomaterials. This text will attempt to bridge the gap and bring together the various properties with wide breadth and scope.

New physical properties and new technologies both in sample preparation and device fabrication evoke on account of the development of nanoscience. Various research fields including physics, chemists, material scientists, and engineers of mechanical and electrical are involved in this research. Nanomaterials are not simply another step in miniaturization, but a different arena entirely; the nanoworld lies midway between the scale of atomic and quantum phenomena, and the scale of bulk materials. At the nanomaterial level, some material properties are affected by the laws of atomic physics, rather than behaving as traditional bulk materials do. The volume discusses the exotic physical properties concerning the linear and nonlinear optical spectra, temperature dependence of resistivities, spin resonance spectra, and magnetic susceptibility measurements. A number of fascinating and provocative results have been developed that lead our perspective understanding of quantum tunnelling, quantum phase transition, surface effect, quantum size-effect confinement and nonlinear susceptibility enhancements. Two key factors cause the properties of nanomaterials to be special: their quantum effects and their structure. Their tiny structure means they have a greater relative surface area than other materials and this can alter or improve properties such as strength and electrical characteristics or reactivity. The properties of materials change as their size approaches the nanoscale and as the percentage of atoms at the surface of a material becomes significant. For bulk materials larger than one micrometre the percentage of atoms at the surface is minuscule relative to the total number of atoms of the material. The interesting and sometimes unexpected properties of nanoparticles are not partly due to the aspects of the surface of the material dominating the properties in lieu of the bulk properties. Their quantum effect can affect the electrical, magnetic or optical performance. The volume discusses various properties of nanomaterials including mechanical, catalytic, magnetic, optical etc. Nanomaterials are subject to the large amount of grain boundaries in bulk materials made of nanoparticles allows extended grain boundary sliding leading to high plasticity. Due to their large surface, nanoparticles made of transition element oxides exhibit interesting catalytic properties. In special cases, catalysis may be enhanced and more specific by decorating these particles with gold or platinum clusters. The large surface area to volume ratio results in a substantial proportion of atoms having different magnetic coupling with neighbouring atoms leading to differing magnetic properties. In magnetic nanoparticles, the energy of magnetic anisotropy may be that small that the vector of magnetization fluctuates thermally; this is called superparamagnetism. Magnetic nanoparticles are used in a range of applications like imaging, bioprocessing, refrigeration as well as high storage density magnetic memory media. Distributions of non-agglomerated nanoparticles in a polymer are used to tune the index of refraction. Additionally, such a process may produce materials with non-linear optical properties. The physical properties of nanomaterials, comprising structure, electrical and magnetic characters, linear optical and nonlinear optical properties subjecting to classical size and quantum size effects are succinctly described.

Nanotechnologies are the design, characterization, production, and application of structures, devices, and systems by controlling shape and size at the nanoscale. Nanomaterials cross the boundary between nanoscience and nanotechnologies and link these both areas together. Generally, nanomaterials deal with sizes of 100 nanometers or smaller in at least one dimension. The material properties of nanostructures are different from the bulk due to the high surface area over volume ratio and possible appearance of quantum effects at the nanoscale. The study of size and shape effects on material properties has attracted enormous attention due to their scientific and industrial importance. This guide to the properties of nanomaterials and nanotechnology in particular addresses the needs of engineers who need to know the special phenomena and potentials, without getting bogged down in the scientific detail of the physics and chemistry involved.

Dr. Rita Khare

1

Introduction

Particle Synthesis by Chemical Routes

Nanomaterials are cornerstones of nanoscience and nanotechnology. Nanostructure science and technology is a broad and interdisciplinary area of research and development activity that has been growing explosively worldwide in the past few years. It has the potential for revolutionizing the ways in which materials and products are created and the range and nature of functionalities that can be accessed. It is already having a significant commercial impact, which will assuredly increase in the future. Some nanomaterials occur naturally, but of particular interest are engineered nanomaterials (EN), which are designed for, and already being used in many commercial products and processes. They can be found in such things as sunscreens, cosmetics, sporting goods, stain-resistant clothing, tires, electronics, as well as many other everyday items, and are used in medicine for purposes of diagnosis, imaging and drug delivery. Engineered nanomaterials are resources designed at the molecular nanometre) level to take advantage of their small size and novel properties which are generally not seen in their conventional, bulk counterparts. The two main reasons why materials at the nano scale can have different properties are increased relative surface area and new quantum effects. Nanomaterials have a much greater surface area to volume ratio than their conventional forms, which can lead to greater chemical reactivity and affect their strength. Also at the nano scale, quantum effects can become much more important in determining the materials properties and characteristics, leading to novel optical, electrical and magnetic behaviours.

Nanomaterials are already in commercial use, with some having been available for several years or decades.

The range of commercial products available today is very broad, including stain-resistant and wrinkle-free textiles, cosmetics, sunscreens, electronics, paints and varnishes. Nanocoatings and nanocomposites are finding uses in book title diverse consumer products, such as windows, sports equipment, bicycles and automobiles. There are novel UV-blocking coatings on glass bottles which protect beverages from damage by sunlight, and longer-lasting tennis balls using butyl-rubber/nano-clay composites. Nanoscale titanium dioxide, for instance, is finding applications in cosmetics, sun-block creams and self-cleaning windows, and nanoscale silica is being used as filler in a range of products, including cosmetics and dental fillings.

The history of nanomaterials began immediately after the big bang when Nanostructures were formed in the early meteorites. Nature later evolved many other Nanostructures like seashells, skeletons etc. Nanoscaled smoke particles were formed during the use of fire by early humans. The scientific story of nanomaterials however began much later. One of the first scientific report is the colloidal gold particles synthesised by Michael Faraday as early as 1857. Nanostructured catalysts have also been investigated for over 70 years. By the early 1940's, precipitated and fumed silica nanoparticles were being manufactured and sold in USA and Germany as substitutes for ultrafine carbon black for rubber reinforcements.

Nanosized amorphous silica particles have found large-scale applications in many every-day consumer products, ranging from non-diary coffee creamer to automobile tires, optical fibres and catalyst supports. In the 1960s and 1970's metallic nanopowders for magnetic recording tapes were developed.

In 1976, for the first time, nanocrystals produced by the now popular inert-gas evaporation technique was published by Granqvist and Buhrman. Recently it has been found that the Maya blue paint is a nanostructured hybrid material. The origin of its colour and its resistance to acids and biocorrosion are still not understood but studies of authentic samples from Jaina Island show that the material is made of needle-shaped palygorskite (clay) crystals that form a superlattice with a period of 1.4 nm, with intercalates of amorphous silicate substrate containing inclusions of metal (Mg) nanoparticles. The beautiful tone of the blue colour is obtained only when both these nanoparticles a nd the superlattice are present, as has been shown by the fabrication of synthetic samples.

Today nanophase engineering expands in a rapidly growing number of structural and functional materials, both inorganic and organic, allowing to manipulate mechanical, catalytic, electric, magnetic, optical and electronic functions. The production of nanophase or cluster-assembled materials is usually based upon the creation of separated small clusters which then are fused into a bulk-like material or on their embedding into compact liquid or solid matrix materials. e.g. nanophase silicon, which differs from normal silicon in physical and electronic properties, could be applied to macroscopic semiconductor processes to create new devices. For instance, when ordinary glass is doped with quantized semiconductor "colloids, " it becomes a high performance optical medium with potential applications in optical computing.

Classification of Nanomaterials

Nanomaterials have extremely small size which having at least one dimension 100 nm or less. Nanomaterials can be nanoscale in one dimension (eg. surface films), two dimensions (eg. strands or fibres), or three dimensions (eg. particles). They can exist in single, fused, aggregated or agglomerated forms with spherical, tubular, and irregular shapes. Common types of nanomaterials include nanotubes, dendrimers, quantum dots and fullerenes.

Nanomaterials have applications in the field of nano technology, and displays different physical chemical characteristics from normal chemicals (i.e., silver nano, carbon nanotube, fullerene, photocatalyst, carbon nano, silica). According to Siegel, Nanostructured material s are classified as Zero dimensional, one dimensional, two dimensional, three dimensional nanostructures.

Nanomaterials are materials which are characterized by an ultra fine grain size (< 50 nm) or by a dimensionality limited to 50 nm. Nanomaterials can be created with various modulation dimensionalities as defined by Richard W. Siegel: zero (atomic clusters, filaments and cluster assemblies), one (multilayers), two (ultrafine-grained overlayers or buried layers), and three (nanophase materials consisting of equiaxed nanometre sized grains).

These materials have created a high interest in recent years by virtue of their unusual mechanical, electrical, optical and magnetic properties. Some examples are given below:

(i) Nanophase ceramics are of particular interest because they are more ductile at elevated temperatures as compared to the coarse-grained ceramics.

(ii) Nanostructured semiconductors are known to show various non-linear optical properties. Semiconductor Q-particles also show quantum confinement effects which may lead to special properties, like the luminescence in silicon powders and silicon germanium quantum dots as infrared optoelectronic devices. Nanostructured semiconductors are used as window layers in solar cells.

(iii) Nanosized metallic powders have been used for the production of gas tight materials, dense parts and porous coatings. Cold welding properties combined with the ductility make them suitable for metal-metal bonding especially in the electronic industry.

(iv) Single nanosized magnetic particles are mono-domains and one expects that also in magnetic nanophase materials the grains correspond with domains, while boundaries on the contrary to disordered walls. Very small particles have special atomic structures with discrete electronic states, which give rise to special properties in addition to the superparamagnetism behaviour. Magnetic nanocomposites have been used for mechanical force transfer (ferrofluids), for high density information storage and magnetic refrigeration. (v) Nanostructured metal clusters and colloids of mono- or plurimetallic composition have a special impact in catalytic applications. They may serve as precursors for new type of heterogeneous catalysts (Cortex-catalysts) and have been shown to offer substantial advantages concerning activity, selectivity and lifetime in chemical transformations and electrocatalysis (fuel cells). Enantioselective catalysis was also achieved using chiral modifiers on the surface of nanoscale metal particles.

(vi) Nanostructured metal-oxide thin films are receiving a growing attention for the realisation of gas sensors (NOx, CO, CO2, CH4 and aromatic hydrocarbons) with enhanced sensitivity and selectivity. Nanostructured metal-oxide (MnO2) finds application for rechargeable batteries for cars or consumer goods. Nanocrystalline silicon films for highly transparent contacts in thin film solar cell and nano-structured titanium oxide porous films for its high transmission and significant surface area enhancement leading to strong absorption indye sensitized solar cells.

(vii) Polymer based composites with a high content of inorganic particles leading to a high dielectric constant are interesting materials for photonic band gap structure.

Examples of Nanomaterials

Nanomaterials (gold, carbon, metals, meta oxides and alloys) with variety of Morphologies.

There are many different ways of creating nanostructures: of course, macromolecules or nanoparticles or buckyballs ornanotubes and so on can be synthesized artificially for certain specific materials. They can also be arranged by methods based on equilibrium or near-equilibrium thermodynamics such as methods of self-organisation and self-assembly (sometimes also called bio-mimetic processes). Using these methods, synthesized materials can be arranged into useful shapes so that finally the material can be applied to a certain application.

Mechanical attrition is a typical example of 'top down' method of synthesis of nanomaterials, where the material is prepared not by cluster assembly but by the structural decomposition of coarser-grained structures as the result of severe plastic deformation. This has become a popular method to make nanocrystalline materials because of its simplicity, the relatively inexpensive equipment needed, and the applicability to essentially the synthesis of all classes of materials. The major advantage often quoted is the possibility for easily scaling up to tonnage quantities of material for various applications. Similarly, the serious problems that are usually cited are;

1. contamination from milling media and/or atmosphere, and
2. to consolidate the powder product without coarsening the nanocrystalline microstructure.

In fact, the contamination problem is often given as a reason to dismiss the method, at least for some materials. Here we will review the mechanisms presently believed responsible for formation of nanocrystalline structures by mechanical attrition of single phase powders, mechanical alloying of dissimilar powders, and mechanical crystallisation of amorphous materials. The two important problems of contamination and powder consolidation will be briefly considered.

Mechanical milling is typically achieved using high energy shaker, planetary ball, or tumbler mills. The energy transferred to the powder from refractory or steel balls depends on the rotational (vibrational) speed, size and number of the balls, ratio of the ball to powder mass, the time of milling and the milling atmosphere. Nanoparticles are produced by the shear action during grinding.

Milling in cryogenic liquids can greatly increase the brittleness of the powders influencing the fracture process. As with any process that produces fine particles, an adequate step to prevent oxidation is necessary. Hence this process is very restrictive for the production of non-oxide materials since then it requires that the milling take place in an inert atmosphere and that the powder particles be handled in an appropriate vacuum system or glove box. This method of synthesis is suitable for producing amorphous or

nanocrystalline alloy particles, elemental or compound powders. If the mechanical milling imparts sufficient energy to the constituent powders a homogeneous alloy can be formed. Based on the energy of the milling process and thermodynamic properties of the constituents the alloy can be rendered amorphous by this processing.

Wet Chemical Synthesis of Nanomaterials

In principle we can classify the wet chemical synthesis of nanomaterials into two broad groups:

1. The top down method: where single crystals are etched in an aqueous solution for producing nanomaterials, For example, the synthesis of porous silicon by electrochemical etching.
2. The bottom up method: consisting of sol-gel method, precipitation etc. where materials containing the desired precursors are mixed in a controlled fashion to form a colloidal solution.

Sol-Gel Process

The sol-gel process, involves the evolution of inorganic networks through the formation of a colloidal suspension (sol) and gelation of the sol to form a network in a continuous liquid phase (gel). The precursors for synthesizing these colloids consist usually of a metal or metalloid element surrounded by various reactive ligands. The starting material is processed to form a dispersible oxide and forms a sol in contact with water or dilute acid. Removal of the liquid from the sol yields the gel, and the sol/gel transition controls the particle size and shape. Calcination of the gel produces the oxide.

Processing of Nanostructured Solgel Materials

In materials science, the sol-gel process is a method for producing solid materials from small molecules. The method is used for the fabrication of metal oxides, especially the oxides of silicon and titanium. The process involves conversion of monomers into a colloidal solution (*sol*) that acts as the precursor for an integrated network (or *gel*) of either discrete particles or network polymers. Typical precursors are metal alkoxides.

Stages in the Process

In this chemical procedure, the 'sol' (or solution) gradually evolves towards the formation of a gel-like diphasic system containing both a liquid phase and solid phase whose morphologies range from discrete particles to continuous polymer networks. In the case of the colloid, the volume fraction of particles (or particle density) may be so low that a significant amount of fluid may need to be removed initially for the gel-like properties to be recognised. This can be accomplished in any number of ways. The simplest method is to allow time for sedimentation to occur, and then pour off the remaining liquid. Centrifugation can also be used to accelerate the process of phase separation.

Removal of the remaining liquid (solvent) phase requires a drying process, which is typically accompanied by a significant amount of shrinkage and densification. The rate at which the solvent can be removed is ultimately determined by the distribution of porosity in the gel. The ultimate microstructure of the final component will clearly be strongly influenced by changes imposed upon the structural template during this phase of processing. Afterwards, a thermal treatment, or firing process, is often necessary in order to favour further polycondensation and enhance mechanical properties and structural stability via final sintering, densification and grain growth. One of the distinct advantages of using this methodology as opposed to the more traditional processing techniques is that densification is often achieved at a much lower temperature.

The precursor sol can be either deposited on a substrate to form a film (e.g., by dip coating or spin coating), cast into a suitable container with the desired shape (e.g., to obtain monolithic ceramics, glasses, fibres, membranes, aerogels), or used to synthesize powders (e.g., microspheres, nanospheres). The sol-gel approach is a cheap and low-temperature technique that allows for the fine control of the product's chemical composition. Even small quantities of dopants, such as organic dyes and rare earth elements, can be introduced in the sol and end up uniformly dispersed in the final product. It can be used in ceramics processing and manufacturing as an investment casting material, or as a means of producing very thin films of metal oxides for various purposes. Sol-gel derived materials have diverse applications in optics, electronics, energy, space, (bio)sensors, medicine (e.g., controlled drug release), reactive material and separation (e.g., chromatography) technology.

The interest in sol-gel processing can be traced back in the mid-1800s with the observation that the hydrolysis of tetraethyl orthosilicate (TEOS) under acidic conditions led to the formation of SiO_2 in the form of fibres and monoliths. Sol-gel research grew to be so important that in the 1990s more than 35,000 papers were published worldwide on the process.

Particles and Polymers

The sol-gel process is a wet-chemical technique used for the fabrication of both glassy and ceramic materials. In this process, the sol (or solution) evolves gradually towards the formation of a gel-like network containing both a liquid phase and a solid phase. Typical precursors are metal alkoxides and metal chlorides, which undergo hydrolysis and polycondensation reactions to form a colloid. The basic structure or morphology of the solid phase can range anywhere from discrete colloidal particles to continuous chain-like polymer networks.

The term *colloid* is used primarily to describe a broad range of solid-liquid (and/or liquid-liquid) mixtures, all of which contain distinct solid (and/or liquid) particles which are dispersed to various degrees in a liquid medium. The term is specific to the size of the individual particles, which are larger than atomic dimensions but small enough to exhibit Brownian motion. If the particles are large enough, then their dynamic behaviour in any given period of time in suspension would be governed by forces of gravity and sedimentation. But if they are small enough to be colloids, then their irregular motion in suspension can be attributed to the collective bombardment of a myriad of thermally agitated molecules in the liquid suspending medium, as described originally by Albert Einstein in his dissertation. Einstein concluded that this erratic behaviour could adequately be described using the theory of Brownian motion, with sedimentation being a possible long term result. This critical size range (or particle diameter) typically ranges from tens of angstroms (10^{-10} m) to a few micrometres (10^{-6} m).

- Under certain chemical conditions (typically in base-catalyzed sols), the particles may grow to sufficient size to become colloids, which are affected both by sedimentation and forces of gravity. Stabilized suspensions of such sub-micrometre spherical particles may eventually result in their self-assembly—yielding highly ordered microstructures reminiscent of the prototype colloidal crystal: precious opal.

- Under certain chemical conditions (typically in acid-catalyzed sols), the interparticle forces have sufficient strength to cause considerable aggregation and/or flocculation prior to their growth. The formation of a more open continuous network of low density polymers exhibits certain advantages with regard to physical properties in the formation of high performance glass and glass/ceramic components in 2 and 3 dimensions.

In either case (discrete particles or continuous polymer network) the sol evolves then towards the formation of an inorganic network containing a liquid phase (gel). Formation of a metal oxide involves connecting the metal centres with oxo (M-O-M) or hydroxo (M-OH-M) bridges, therefore generating metal-oxo or metal-hydroxo polymers in solution.

In both cases (discrete particles or continuous polymer network), the drying process serves to remove the liquid phase from the gel, yielding a micro-porous amorphous glass or micro-crystalline ceramic. Subsequent thermal treatment (firing) may be performed in order to favour further polycondensation and enhance mechanical properties. With the viscosity of a sol adjusted into a proper range, both optical quality glass fibre and refractory ceramic fibre can be drawn which are used for fibre optic sensors and thermal insulation, respectively. In addition, uniform ceramic powders of a wide range of chemical composition can be formed by precipitation.

Polymerization

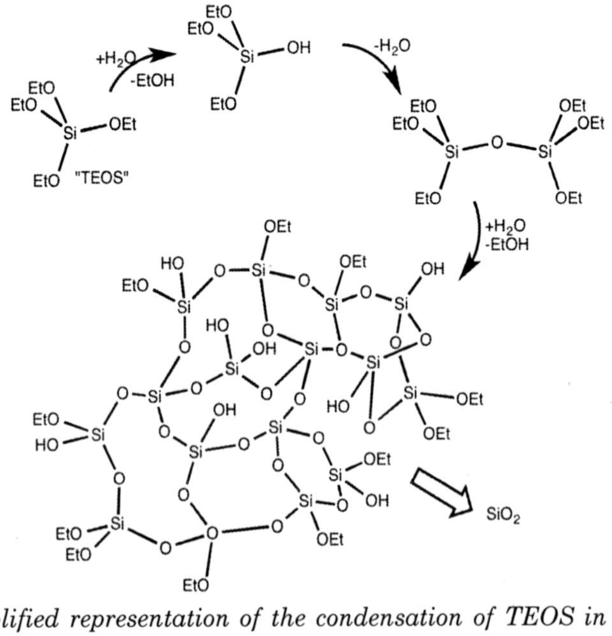

Figure: *Simplified representation of the condensation of TEOS in sol gel process*

A well studied alkoxide is silicon tetraethoxide, or tetraethyl orthosilicate (TEOS). The chemical formula for TEOS is given by $Si(OC_2H_5)_4$, or $Si(OR)_4$, where the alkyl group $R = C_2H_5$. Alkoxides are ideal chemical precursors for sol-gel synthesis because they react readily with water. The reaction is called hydrolysis, because a hydroxyl ion becomes attached to the silicon atom as follows:

$$Si(OR)_4 + H_2O \rightarrow HO\text{-}Si(OR)_3 + R\text{-}OH$$

Depending on the amount of water and catalyst present, hydrolysis may proceed to completion to silica:

$$Si(OR)_4 + 2\ H_2O \rightarrow SiO_2 + 4\ R\text{-}OH$$

Complete hydrolysis requires a significant excess of water and catalysts such as acetic acid or hydrochloric acid. Intermediate species include $[(OR)_2\text{–}Si\text{-}(OH)_2]$ or $[(OR)_3\text{–}Si\text{-}(OH)]$, resulting from partial hydrolysis reactions. Early intermediates result from two partially hydrolyzed monomers linked via a siloxane [Si–O–Si] bond:

$$(OR)_3\text{–}Si\text{-}OH + HO\text{–}Si\text{-}(OR)_3 \rightarrow [(OR)_3Si\text{–}O\text{–}Si(OR)_3] + H\text{-}O\text{-}H$$

or

$$(OR)_3\text{–}Si\text{-}OR + HO\text{–}Si\text{-}(OR)_3 \rightarrow [(OR)_3Si\text{–}O\text{–}Si(OR)_3] + R\text{-}OH$$

Thus, polymerization is associated with the formation of a 1-, 2-, or 3-dimensional network of siloxane [Si–O–Si] bonds accompanied by the production of H-O-H and R-O-H species.

By definition, condensation liberates a small molecule, such as water or alcohol. This type of reaction can continue to build larger and larger silicon-containing molecules by the process of polymerization. Thus, a polymer is a huge molecule (or macromolecule) formed from hundreds or thousands of units called monomers. The number of bonds that a monomer can form is called its functionality. Polymerization of silicon alkoxide, for instance, can lead to complex branching of the polymer, because a fully hydrolyzed monomer $Si(OH)_4$ is tetrafunctional (can branch or bond in 4 different directions). Alternatively, under certain conditions (e.g., low water concentration) fewer than 4 of the OR or OH groups (ligands) will be capable of condensation, so relatively little branching will occur. The mechanisms of hydrolysis and condensation, and the factors that bias the structure toward linear or branched structures are the most critical issues of sol-gel science and technology. This reaction is favoured in both basic and acidic conditions.

Sono-Ormosil

Sonication is an efficient tool for the synthesis of polymers. The cavitational shear forces, which stretch out and break the chain in a non-random process, result in a lowering of the molecular weight and poly-dispersity. Furthermore, multi-phase systems are very efficient dispersed and emulsified, so that very fine mixtures are provided. This means that ultrasound increases the rate of polymerisation over conventional stirring and results in higher molecular weights with lower polydispersities. Ormosils (organically modified silicate) are obtained when silane is added to gel-derived silica during sol-gel process. The product is a molecular-scale composite with improved mechanical properties. Sono-Ormosils are characterized by a higher density than classic gels as well as an improved thermal stability. An explanation therefore might be the increased degree of polymerization.

Nanomaterials

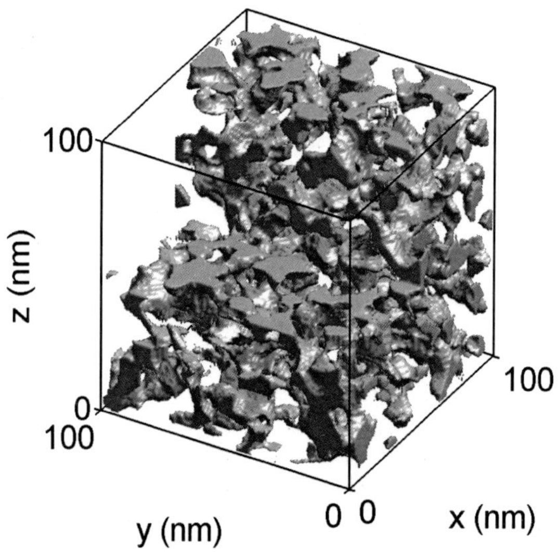

Figure: *Nanostructure of a resorcinol-formaldehyde gel reconstructed from Small-Angle X-ray Scattering. This type of disordered morphology is typical of many sol-gel materials.*

In the processing of fine ceramics, the irregular particle sizes and shapes in a typical powder often lead to non-uniform packing morphologies that result in packing density variations in the powder compact. Uncontrolled flocculation of powders due to attractive van der Waals forces can also give rise to microstructural inhomogeneities.

Differential stresses that develop as a result of non-uniform drying shrinkage are directly related to the rate at which the solvent can be removed, and thus highly dependent upon the distribution of porosity. Such stresses have been associated with a plastic-to-brittle transition in consolidated bodies, and can yield to crack propagation in the unfired body if not relieved.

In addition, any fluctuations in packing density in the compact as it is prepared for the kiln are often amplified during the sintering process, yielding inhomogeneous densification. Some pores and other structural defects associated with density variations have been shown to play a detrimental role in the sintering process by growing and thus limiting end-point densities. Differential stresses arising from inhomogeneous densification have also been shown to result in the propagation of internal cracks, thus becoming the strength-controlling flaws. It would therefore appear desirable to process a material in such a way that it is physically uniform with regard to the distribution of components and porosity, rather than using particle size distributions which will maximize the green density. The containment of a uniformly dispersed assembly of strongly interacting particles in suspension requires total control over particle-particle interactions. Monodisperse colloids provide this potential.

Monodisperse powders of colloidal silica, for example, may therefore be stabilized sufficiently to ensure a high degree of order in the colloidal crystal or polycrystalline colloidal solid which results from aggregation. The degree of order appears to be limited by the time and space allowed for longer-range correlations to be established. Such defective polycrystalline structures would appear to be the basic elements of nanoscale materials science, and, therefore, provide the first step in developing a more rigorous understanding of the mechanisms involved in microstructural evolution in inorganic systems such as sintered ceramic nanomaterials.

Applications

Protective Coatings: The applications for sol gel-derived products are numerous. For example, scientists have used it to produce the world's lightest materials and also some of its toughest ceramics. One of the largest application areas is thin films, which can be produced on a piece of substrate by spin coating or dip coating. Protective and decorative coatings, and electro-optic components can be applied to glass, metal and other types of substrates with these methods. Cast into a mold, and with further drying and heat-treatment, dense ceramic or glass articles with novel properties can be formed that cannot be created by any other method. Other coating methods include spraying, electrophoresis, inkjet printing or roll coating.

Thin Films and Fibres

With the viscosity of a sol adjusted into a proper range, both optical and refractory ceramic fibres can be drawn which are used for fibre optic sensors and thermal insulation, respectively. Thus, many ceramic materials, both glassy and crystalline, have found use in various forms from bulk solid-state components to high surface area forms such as thin films, coatings and fibres.

Nanoscale Powders

Ultra-fine and uniform ceramic powders can be formed by precipitation. These powders of single and multiple component compositions can be produced on a nanoscale particle size for dental and biomedical applications. Composite powders have been patented for use as agrochemicals and herbicides. Powder abrasives, used in a variety of finishing operations, are made using a sol-gel type process. One of the more important applications of sol-gel processing is to carry out zeolite synthesis. Other elements (metals, metal oxides) can be easily incorporated into the final product and the silicate sol formed by this method is very stable. Another application in research is to entrap biomolecules for sensory (biosensors) or catalytic purposes, by physically or chemically preventing them from leaching out and, in the case of protein or chemically-linked small molecules, by shielding them from the external environment yet allowing small molecules to be monitored. The major disadvantages are that the change in local environment may alter the functionality of the protein or small molecule entrapped and that the synthesis step may damage the protein. To circumvent this, various strategies have been explored, such as monomers with protein friendly leaving groups (e.g. glycerol) and the inclusion of polymers which stabilize protein (e.g. PEG).

Other products fabricated with this process include various ceramic membranes for microfiltration, ultrafiltration, nanofiltration, pervaporation and reverse osmosis. If the liquid in a wet gel is removed under a supercritical condition, a highly porous and extremely low density material called aerogel is obtained. Drying the gel by means of low temperature treatments (25-100 °C), it is possible to obtain porous solid matrices called xerogels. In addition, a sol-gel process was developed in the 1950s for the production of radioactive powders of UO_2 and ThO_2 for nuclear fuels, without generation of large quantities of dust.

Opto-Mechanical

Macroscopic optical elements and active optical components as well as large area hot mirrors, cold mirrors, lenses and beam splitters all with optimal geometry can be made quickly and at low cost via the sol-gel route. In the processing of high performance ceramic nanomaterials with superior opto-mechanical properties under adverse conditions, the size of the crystalline grains is determined largely by the size of the crystalline particles present in the raw material during the synthesis or formation of the object. Thus a reduction of the original particle size well below the wavelength of visible light (~500 nm) eliminates much of the light scattering, resulting in a translucent or even transparent material. Furthermore, results indicate that microscopic pores in sintered ceramic nanomaterials, mainly trapped at the junctions of microcrystalline grains, cause light to scatter and prevented true transparency. It has been observed that the total volume fraction of these nanoscale pores (both intergranular and intragranular porosity) must be less than 1% for high-quality optical transmission. I.E. The density has to be 99.99% of the theoretical crystalline density.

Mechanics of Gelation

In a static sense, the fundamental difference between a liquid and a solid is that the solid has elastic resistance against a shearing stress while a liquid does not. Thus, a simple liquid will not typically support a transverse acoustic phonon, or shear wave. Gels have been described by Born as liquids in which an elastic resistance against shearing survives, yielding both viscous and elastic properties. It has been shown theoretically that in a certain low-frequency range, polymeric gels should propagate shear waves with relatively low damping. The distinction between a sol (solution) and a gel therefore appears to be understood in a manner analogous to the practical distinction between the elastic and plastic deformation ranges of a metal. The distinction lies in the ability to respond to an applied shear force via macroscopic viscous flow.

In a dynamic sense, the response of a gel to an alternating force (oscillation or vibration) will depend upon the period or frequency of vibration. As indicated here, even most simple liquids will exhibit some elastic response at shear rates or frequencies exceeding 5 x 10^6 cycles per second. Experiments on such short time scales probe the fundamental motions of the primary particles (or particle clusters) which constitute the lattice structure or aggregate. The increasing resistance of certain liquids to flow at high stirring speeds is one manifestation of this phenomenon. The ability of a condensed body to respond to a mechanical force by viscous flow is thus strongly dependent on the time scale over which the load is applied, and thus the frequency and amplitude of the stress wave in oscillatory experiments.

Structural Relaxation

The structural relaxation of a viscoelastic gel has been identified as primary mechanism responsible for densification and associated pore evolution in both colloidal and polymeric silica gels. Experiments in the viscoelastic properties of such skeletal networks on various time scales require a force varying with a period (or frequency) appropriate to the relaxation time of the phenomenon investigated, and inversely proportional to the distance over which such relaxation occurs. High frequencies associated with ultrasonic waves have been used extensively in the handling of polymer solutions, liquids and gels and the determination of their viscoelastic properties. Static measurements of the shear modulus have been made, as well as dynamic measurements of the speed of propagation of shear waves, a measurement which then yields the dynamic modulus of rigidity. Dynamic Light Scattering (DLS) techniques have been utilised in order to monitor the dynamics of density fluctuations through the behaviour of the autocorrelation function near the point of gelation.

Phase Transition

Tanaka, *et al.*, emphasize that the discrete and reversible volume transitions which occur in partially hydrolyzed acrylimide gels can be interpreted in terms of a phase transition of the system consisting of the charged polymer network, hydrogen (counter)ions and liquid matrix. The phase transition is a manifestation of competition among the three forces which contribute to the osmotic pressure in the gel:

1. The positive osmotic pressure of (+) hydrogen ions
2. The negative pressure due to polymer-polymer affinity
3. The rubber-like elasticity of the polymer network

The balance of these forces varies with change in temperature or solvent properties. The total osmotic pressure acting on the system is the sum osmotic pressure of the gel. It is further shown that the phase transition can be induced by the application of an electrical field across the gel. The volume change at the transition point is either discrete (as in a first-order Ehrenfest transition) or continuous (second order Ehrenfest analogy), depending on the degree of ionization of the gel and on the solvent composition.

Elastic Continuum

The gel is thus interpreted as an elastic continuum which deforms when subjected to externally applied shear forces, but is incompressible upon application of hydrostatic pressure. This combination of fluidity and rigidity is explained in terms of the gel structure: that of a liquid contained within a fibrous polymer network or matrix by the extremely large friction between the liquid and the fibre or polymer network. Thermal fluctuations may produce infinitesimal expansion or contraction within the network, and the evolution of such fluctuations will ultimately determine the molecular morphology and the degree of hydration of the body.

Quasi-elastic light scattering offers direct experimental access to measurement of the wavelength and lifetimes of critical fluctuations, which are governed by the viscoelastic properties of the gel. It is reasonable to expect a relationship between the amplitude of such fluctuations and the elasticity of the network. Since the elasticity measures the resistance of the network to either elastic (reversible) or plastic (irreversible) deformation, the fluctuations should grow larger as the elasticity declines. The divergence of the scattered light intensity at a finite critical temperature implies that the elasticity approaches zero, or the compressibility becomes infinite, which is the typically observed behaviour of a system at the point of instability. Thus, at the critical point, the polymer network offers no resistance at all to any form of deformation.

Ultimate Microstructure

The rate of relaxation of density fluctuations will be rapid if the restoring force, which depends upon the network elasticity, is large—and if the friction between the network and the interstitial fluid is small. The theory suggests that the rate is directly proportional to the elasticity and inversely proportional to the frictional force. The friction in turn depends upon both the viscosity of the fluid and the average size of the pores contained within the polymer network.

Thus, if the elasticity is inferred from the measurements of the scattering intensity, and the viscosity is determined independently (via mechanical methods such as ultrasonic attenuation) measurement of the relaxation rate yields information on the pore size distribution contained within the polymer network, *e.g* large fluctuations in polymer density near the critical point yield large density differentials with a corresponding bimodal distribution of porosity. The difference in average size between the smaller pores (in the highly dense regions) and the larger pores (in regions of lower average density) will therefore depend upon the degree of phase separation which is allowed to occur before such fluctuations become thermally arrested or "frozen in" at or near the critical point of the transition.

Chemical Properties

A chemical property is any of a material's properties that becomes evident during a chemical reaction; that is, any quality that can be established only by changing a substance's chemical identity. Simply speaking, chemical properties cannot be determined just by viewing or touching the substance; the substance's internal structure must be affected for its chemical properties to be investigated. However a catalytic property would also be a chemical property.

Chemical properties can be contrasted with physical properties, which can be discerned without changing the substance's structure. However, for many properties within the scope of physical chemistry, and other disciplines at the boundary between chemistry and physics, the distinction may be a matter of researcher's perspective. Material properties, both physical and chemical, can be viewed as supervenient; i.e., secondary to the underlying reality. Several layers of superveniency are possible.

Chemical properties can be used for building chemical classifications. They can also be useful to identify an unknown substance or to separate or purify it from other substances. Materials science will normally consider the chemical properties of a substance to guide its applications.

Examples of Chemical Properties

Heat of Combustion

The heat of combustion (ΔH_c°) is the energy released as heat when a compound undergoes complete combustion with oxygen under standard conditions. The chemical reaction is typically a hydrocarbon reacting with oxygen to form carbon dioxide, water and heat. It may be expressed with the quantities:

- energy/mole of fuel (kJ/mol)
- energy/mass of fuel
- energy/volume of fuel

The heat of combustion is conventionally measured with a bomb calorimetre. It may also be calculated as the difference between the heat of formation ΔH_f° of the products and reactants.

Heating Value

The heating value (or energy value or calorific value) of a substance, usually a fuel or food, is the amount of heat released during the combustion of a specified amount of it. The energy value is a characteristic for each substance. It is measured in units of energy per unit of the substance, usually mass, such as: kJ/kg, kJ/mol, kcal/kg, Btu/lb. Heating value is commonly determined by use of a bomb calorimetre.

Heating value unit conversions (for more visit Wolfram Alpha):

- kcal/kg = MJ/kg * 238.846
- Btu/lb = MJ/kg * 429.923
- Btu/lb = kcals * 1.8

The heat of combustion for fuels is expressed as the HHV, LHV, or GHV.

Higher Heating Value

The quantity known as higher heating value (HHV) (or *gross energy* or *upper heating value* or *gross calorific value* (GCV) or *higher calorific value* (HCV)) is determined by bringing all the products of combustion back to the original pre-combustion temperature, and in particular condensing any vapor produced. Such measurements often use a standard temperature of 25 °C. This is the same as the thermodynamic heat of combustion since the enthalpy change for the reaction assumes a common temperature of the compounds before and after combustion, in which case the water produced by combustion is liquid.

The higher heating value takes into account the latent heat of vaporization of water in the combustion products, and is useful in calculating heating values for fuels where condensation of the reaction products is practical (e.g., in a gas-fired boiler used for space heat). In other words, HHV assumes all the water component is in liquid state at the end of combustion (in product of combustion) and that heat below 150 °C can be put to use.

Lower Heating Value

The quantity known as lower heating value (LHV) (*net calorific value* (NCV) or *lower calorific value* (LCV)) is determined by subtracting the heat of vaporization of the water vapor from the higher heating value. This treats any H_2O formed as a vapor. The energy required to vaporize the water therefore is not released as heat.

LHV calculations assume that the water component of a combustion process is in vapor state at the end of combustion, as opposed to the higher heating value (HHV) (a.k.a. *gross calorific value* or *gross CV*) which assumes that all of the water in a combustion process is in a liquid state after a combustion process. The LHV assumes that the latent heat of vaporization of water in the fuel and the reaction products is not recovered. It is useful in comparing fuels where condensation of the combustion products is impractical, or heat at a temperature below 150 °C cannot be put to use.

The above is but one definition of lower heating value adopted by the American Petroleum Institute (API) and uses a reference temperature of 60 °F (15.56 °C).

Another definition, used by Gas Processors Suppliers Association (GPSA) and originally used by API (data collected for API research project 44), is the enthalpy of all combustion products minus the enthalpy of the fuel at the reference temperature (API research project 44 used 25 °C. GPSA currently uses 60 °F), minus the enthalpy of the stoichiometric oxygen (O_2) at the reference temperature, minus the heat of vaporization of the vapor content of the combustion products.

The distinction between the two is that this second definition assumes that the combustion products are all returned to the reference temperature and the heat content from the condensing vapor is considered not to be useful. This is more easily calculated from the higher heating value than when using the preceding definition and will in fact give a slightly different answer.

Gross Heating Value

- Gross heating value accounts for water in the exhaust leaving as vapor, and includes liquid water in the fuel prior to combustion. This value is important for fuels like wood or coal, which will usually contain some amount of water prior to burning.
- Note that GPSA 12th Edition states that the Gross Heating Value of a gas is equivalent to Higher Heating Value. This suggests that there may be different standards in play. The use of the term Gross normally describes a larger value than the Net, which usually describes a smaller value. GPSA is consistent with this, and equates the Gross Heating Value to the higher heating value (for a gas - so probably with no liquid water present), and the Net Heating Value to the lower heating value.

Measuring Heating Values

The higher heating value is experimentally determined in a bomb calorimetre. The combustion of a stoichiometric mixture of fuel and oxidizer (e.g. two moles of hydrogen and one mole of oxygen) in a steel container at 25 °C is initiated by an ignition device and the reactions allowed to complete. When hydrogen and oxygen react during combustion, water vapor is produced. The vessel and its contents are then cooled to the original 25 °C and the higher heating value is determined as the heat released between identical initial and final temperatures.

When the lower heating value (LHV) is determined, cooling is stopped at 150 °C and the reaction heat is only partially recovered. The limit of 150 °C is an arbitrary choice.

Note: Higher heating value (HHV) is calculated with the product of water being in liquid form while lower heating value (LHV) is calculated with the product of water being in vapor form.

Relation between Heating Values

The difference between the two heating values depends on the chemical composition of the fuel. In the case of pure carbon or carbon monoxide, the two heating values are almost identical, the difference being the sensible heat content of carbon dioxide between 150 °C and 25 °C (sensible heat exchange causes a change of temperature. In contrast, latent heat is added or subtracted for phase transitions at constant temperature. Examples: heat of vaporization or heat of fusion). For hydrogen the difference is much more significant as it includes the sensible heat of water vapor between 150 °C and 100 °C, the latent heat of condensation at 100 °C, and the sensible heat of the condensed water between 100 °C and 25 °C. All in all, the higher heating value of hydrogen is 18.2% above its lower heating value (142 MJ/kg vs. 120 MJ/kg). For hydrocarbons the difference depends on the hydrogen content of the fuel. For gasoline and diesel the higher heating value exceeds the lower heating value by about 10% and 7% respectively, and for natural gas about 11%.

A common method of relating HHV to LHV is:

$$HHV = LHV + h_{\upsilon} \times (n_{H2O,out}/n_{fuel,in})$$

where h_{υ} is the heat of vaporization of water, $n_{H2O,out}$ is the moles of water vaporized and $n_{fuel,in}$ is the number of moles of fuel combusted.

Most applications that burn fuel produce water vapor, which is unused and thus wastes its heat content. In such applications, the lower heating value is generally used to give a 'benchmark' for the process; however, for true energy calculations the higher heating value is correct. This is particularly relevant for natural gas, whose high hydrogen content produces much water. The gross energy value is relevant for gas burned in condensing boilers and power plants with flue-gas condensation that condense the water vapor produced by combustion, recovering heat which would otherwise be wasted.

Usage of terms

For historical reasons the efficiency of power plants and combined heat and power plants in Europe may have once been calculated based on the LHV. However, this does not seem to be the case nowadays and most countries are tending to correctly use HHV for true efficiency figures. This is becoming noticeable in modern UK energy publications with the increase of energy awareness and based on the simple fact that it is correct. While in the US, values may have been reported to be generally based on the HHV, although any initial investigation may reveal that the US is still tending to use LHV in some circumstances, whether technically correct or not. This has the peculiar result that contemporary combined heat and power plants, where flue-gas condensation is implemented, may report efficiencies exceeding 100%. Using LHV in other energy calculations brings similar errors, especially when pulled (incorrectly) into electrolysis calculations etc.

Many engine manufacturers rate their engine fuel consumption by the lower heating values. American consumers should be aware that the corresponding fuel-consumption figure based on the higher heating value will be somewhat higher.

The difference between HHV and LHV definitions causes endless confusion when quoters do not bother to state the convention being used. since there is typically a 10% difference between the two methods for a power plant burning natural gas. For simply benchmarking part of a reaction the LHV may be appropriate, but HHV should be used for overall energy efficiency calculations, if only to avoid confusion, and in any case the value or convention should be clearly stated.

Accounting for Moisture

Both HHV and LHV can be expressed in terms of AR (all moisture counted), MF and MAF (only water from combustion of hydrogen). AR, MF, and MAF are commonly used for indicating the heating values of coal:

- AR (As Received) indicates that the fuel heating value has been measured with all moisture and ash forming minerals present.
- MF (Moisture Free) or Dry indicates that the fuel heating value has been measured after the fuel has been dried of all inherent moisture but still retaining its ash forming minerals.
- MAF (Moisture and Ash Free) or DAF (Dry and Ash Free) indicates that the fuel heating value has been measured in the absence of inherent moisture and ash forming minerals.

Enthalpy of Formation

The standard enthalpy of formation or standard heat of formation of a compound is the change of enthalpy from the formation of 1 mole of the compound from its constituent elements, with all substances in their standard states at 1 atmosphere (1 atm or 101.3 kPa). Its symbol is ΔH_f^{Θ} or $\Delta_f H^{\Theta}$. The superscript theta (zero) on this symbol indicates that the process has occurred under standard conditions at the specified temperature (usually 25 degrees Celsius or 298.15 K). Standard states are as follows:

1. For a gas: the standard state is a pressure of exactly 1 atm
2. For a solute present in an ideal solution: a concentration of exactly 1 M at a pressure of 1 atm
3. For a pure substance or a solvent in a condensed state (a liquid or a solid): the standard state is the pure liquid or solid under a pressure of 1 atm
4. For an element: the form in which the element is most stable under 1 atm of pressure. One exception is phosphorus, for which the most stable form at 1 atm is black phosphorus, but white phosphorus is chosen as the standard reference state for zero enthalpy of formation.

For example, the standard enthalpy of formation of carbon dioxide would be the enthalpy of the following reaction under the conditions above:

$$C_{(s,graphite)} + O_{2(g)} \rightarrow CO_{2(g)}$$

Note that all elements are written in their standard states, and one mole of product is formed. This is true for all enthalpies of formation.

The standard enthalpy of formation is measured in units of energy per amount of substance. Most are defined in kilojoules per mole (kJ mol"1), but can also be measured in calories per mole, joules per mole or kilocalories per gram (any combination of these units conforming to the energy per mass or amount guideline). In physics the energy per particle is often expressed in electronvolts which corresponds to about 100 kJ mol^{-1}.

All elements in their standard states (oxygen gas, solid carbon in the form of graphite, etc.) have a standard enthalpy of formation of zero, as there is no change involved in their formation.

The formation reaction is a constant pressure and constant temperature process. Since the pressure of the standard formation reaction is fixed at 1 atm, the standard formation enthalpy or reaction heat is a function of temperature. For tabulation purposes, standard formation enthalpies are all given at a single temperature: 298 K, represented by the symbol ΔH_{f298}^{Θ} .

Mechanics

The standard enthalpy of formation is equivalent to the sum of many separate processes included in the Born-Haber cycle of synthesis reactions. For example, to calculate the standard enthalpy of formation of sodium chloride, we use the following reaction:

$$Na_{(s)} + (1/2)Cl_{2(g)} \rightarrow NaCl_{(s)}$$

This process is made of many separate sub-processes, each with their own enthalpies. Therefore, we must take into account:

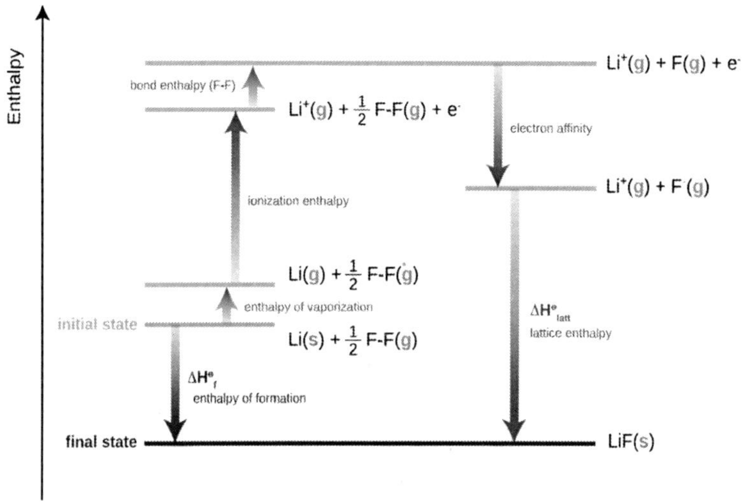

***Figure:** Standard enthalpy change of formation in Born-Haber diagram for lithium fluoride.*

1. The standard enthalpy of atomization of solid sodium
2. The first ionization energy of gaseous sodium
3. The standard enthalpy of atomization of chlorine gas
4. The electron affinity of chlorine atoms
5. The lattice enthalpy of sodium chloride

The sum of all these values will give the standard enthalpy of formation of sodium chloride.

Additionally, applying Hess's Law shows that the *sum* of the individual reactions corresponding to the enthalpy change of formation for each substance in the reaction is equal to the enthalpy change of the overall reaction, regardless of the number of steps or intermediate reactions involved. This is because enthalpy is a state function. In the example above the standard enthalpy change of formation for sodium chloride is equal to the sum of the standard enthalpy change of formation for each of the steps involved in the process. This is especially useful for very long reactions with many intermediate steps and compounds.

Chemists may use standard enthalpies of formation for a reaction that is hypothetical. For instance carbon and hydrogen will not directly react to form methane, yet the standard enthalpy of formation for methane is determined to be -74.8 kJ mol^{-1} from using other known standard enthalpies of reaction with Hess's law. That it is negative shows that the reaction, if it were to proceed, would be exothermic; that is, it is enthalpically more stable than hydrogen gas and carbon.

It is possible to predict heat of formations for simple unstrained organic compounds with the Heat of formation group additivity method.

Standard Enthalpy of Reaction

Standard enthalpies of formation are used in thermochemistry to find the standard enthalpy change of any reaction. This is done by subtracting the sum of the standard enthalpies of formation of the reactants (each being multiplied by its respective stoichiometric coefficient, ν) from the sum of the standard enthalpies of formation of the products (each also multiplied by its respective stoichiometric coefficient), as shown in the equation below:

$$\Delta H^\circ = \Sigma(\nu \times \Delta H_f^\circ) \text{ (products)} - \Sigma(\nu \times \Delta H_f^\circ) \text{ (reactants)}$$

For example, for the reaction $CH_4 + 2\ O_2 \rightarrow CO_2 + 2\ H_2O$:

$\Delta H_r^\circ = [(1 \times \Delta H_{f\ (CO2)}^\circ) + (2 \times \Delta H_{f\ (H2O)}^\circ)]$ (products) - $[(1 \times \Delta H_{f\ (CH4)}^\circ) + (2 \times \Delta H_{f\ (O2)}^\circ)]$ (reactants)

If the standard enthalpy of the products is less than the standard enthalpy of the reactants, the standard enthalpy of reaction will be negative. This implies that the reaction is exothermic. The converse is also true; the standard enthalpy of reaction will be positive for an endothermic reaction.

Key Concepts for Doing Enthalpy Calculations

1. When a reaction is reversed, the magnitude of DH stays the same, but the sign changes.
2. When the balanced equation for a reaction is multiplied by an integer, the corresponding value of DH must be multiplied by that integer as well.
3. The change in enthalpy for a reaction can be calculated from the enthalpies of formation of the reactants and the products
4. Elements in their standard states are not included in the enthalpy calculations for the reaction since the enthalpy of an element in its standard state is zero.

Subcategories

- Standard enthalpy of neutralization is the change in enthalpy that occurs when an acid and base undergo a neutralization reaction to form one mole of water under standard conditions, as previously defined.
- Standard enthalpy of sublimation, or heat of sublimation, is defined as the enthalpy required to sublime one mole of the substance under standard conditions, as previously defined.

- Standard enthalpy of solution (or enthalpy change of dissolution or heat of solution) is the enthalpy change associated with the dissolution of a substance in a solvent at constant pressure under standard conditions, as previously defined.
- Standard enthalpy of hydrogenation is defined as the enthalpy change observed when one mole of an unsaturated compound reacts with an excess of hydrogen to become fully saturated under standard conditions, as previously defined.

Examples: Standard Enthalpies of Formation (at 25°C, 298 K)

Chemical Compound	*Phase (Matter)*	*Chemical Formula*	ΔH_f^0 *in kJ/mol*
Acetone	l	C_3H_6O	-248.4
Acetylene	g	C_2H_2	+227.4
Ammonia (Ammonium Hydroxide)	aq	NH_3 (NH_4OH)	-80.8
Ammonia	g	NH_3	-46.1
Ammonium nitrate	s	NH_4NO_3	-365.6
Benzene	l	C_6H_6	+49.1
Bromine	l	Br_2	0
Bromine	g	Br_2	+31
Bromine	g	Br	+111.9
Hydrogen bromide	g	HBr	-36.3
Calcium	s	Ca	0
Calcium oxide	s	CaO	-634.9
Calcium carbonate	s	$CaCO_3$	-1207.6
Carbon	s	C (graphite)	0
Carbon	s	C (diamond)	+1.88
Carbon monoxide	g	CO	-110.5
Carbon dioxide	g	CO_2	-393.5
Chlorine	g	Cl_2	0
Chlorine	g	Cl	+121.3
Hydrogen chloride	g	HCl	-92.3
Copper(II) sulfate	aq	$CuSO_4$	-769.98
Ethane	g	C_2H_6	-84.68
Ethanol	l	C_2H_5OH	-277.6
Ethylene	g	C_2H_4	+52.4
Fluorine	g	F_2	0
Fluorine	g	F	+79.38
Hydrogen fluoride	g	HF	-273.3
Glucose	s	$C_6H_{12}O_6$	-1273.3
Hydrogen	g	H_2	0
Iodine	s	I_2	0

Contd...

Chemical Compound	*Phase (Matter)*	*Chemical Formula*	ΔH_f^0 *in kJ/mol*
Iodine	g	I_2	+62
Isopropanol	g	C_3H_7OH	-318.1
Methane	g	CH_4	-74.87
Methanol	l	CH_3OH	-238.6
Nitrogen	g	N	0
Nitrogen dioxide	g	NO_2	+33.2
Nitric oxide	g	NO	+91.3
Oxygen	g	0_2	0
Ozone	g	0_3	+142.7
Propane	g	C_3H_8	-103.85
Sodium	s	Na	0
Sodium	g	Na	+107.5
Sodium bicarbonate	s	$NaHCO_3$	-950.8
Sodium carbonate	s	Na_2CO_3	-1131
Sodium chloride (table salt)	aq	NaCl	-407
Sodium chloride (table salt)	s	NaCl	-411.12
Sodium chloride (table salt)	l	NaCl	-385.92
Sodium chloride (table salt)	g	NaCl	-181.42
Sodium hydroxide	aq	NaOH	-470.1
Sodium hydroxide	s	NaOH	-426.7
Sodium nitrate	aq	$NaNO_3$	-446.2
Sodium nitrate	s	$NaNO_3$	-424.8
Sucrose	s	$C_{12}H_{22}0_{11}$	-2226.1
Sulfur (monoclinic)	s	S_8	0.3
Sulfur (rhombic)	s	S_8	0
Sulfur dioxide	g	SO_2	-296.8
Sulfur trioxide	g	SO_3	-395.7
Sulfuric acid	l	H_2SO_4	-814
Silica	s	$Si0_2$	-911
Silver	s	Ag	0
Silver chloride	s	AgCl	-127.0
Water	l	H_20	-285.8
Water vapor	g	H_20	-241.82
Zinc sulfate	s	$ZnSO_4$	-980.14

Toxicity

Toxicity is the degree to which a substance can damage an organism. Toxicity can refer to the effect on a whole organism, such as an animal, bacterium, or plant, as well as the effect on a substructure of the organism, such as a cell (cytotoxicity) or an organ such as the liver (hepatotoxicity). By extension, the word may be metaphorically used to describe toxic effects on larger and more complex groups, such as the family unit or society at large.

A central concept of toxicology is that effects are dose-dependent; even water can lead to water intoxication when taken in too high a dose, whereas for even a very toxic substance such as snake venom there is a dose below which there is no detectable toxic effect. Toxicity is species-specific, lending cross-species analysis problematic. Newer paradigms and metrics are evolving to bypass animal testing, while maintaining the concept of toxicity endpoints.

Types of Toxicity

There are generally three types of toxic entities; chemical, biological, and physical:

- Chemical toxicants include inorganic substances such as lead, mercury, hydrofluoric acid, and chlorine gas, and organic compounds such as methyl alcohol, most medications, and poisons from living things. Radioactive chemicals are not poisonous because of their chemical nature, but because radiations emitted by nuclei are highly energetic, and destroy cells and tissues, but radioactive toxicity should not be comprehended in chemical toxicity.
- Biological toxicants include bacteria and viruses that can induce disease in living organisms. Biological toxicity can be difficult to measure because the "threshold dose" may be a single organism. Theoretically one virus, bacterium or worm can reproduce to cause a serious infection. However, in a host with an intact immune system the inherent toxicity of the organism is balanced by the host's ability to fight back; the effective toxicity is then a combination of both parts of the relationship. A similar situation is also present with other types of toxic agents.
- Physical toxicants are substances that, due to their physical nature, interfere with biological processes. Examples include coal dust, asbestos fibres or finely divided silicon dioxide, all of which can ultimately be fatal if inhaled. Corrosive chemicals possess physical toxicity because they destroy tissues, but they're not directly poisonous unless they interfere directly with biological activity. Water can act as a physical toxicant if taken in extremely high doses because the concentration of vital ions decreases dramatically if there's too much water in the body. Asphyxiant gases can be considered physical toxicants because they act by displacing oxygen in the environment but they are inert, not toxic gases.

Measuring Toxicity

Toxicity can be measured by its effects on the target (organism, organ, tissue or cell). Because individuals typically have different levels of response to the same dose of a toxic substance, a population-level measure of toxicity is often used which relates the probabilities of an outcome for a given individual in a population. One such measure is the LD_{50}. When such data does not exist, estimates are made by comparison to known similar toxic things, or to similar exposures in similar organisms. Then, "safety factors" are added to account for uncertainties in data and evaluation processes. For example, if a dose of a toxic substance is safe for a laboratory rat, one might assume that one tenth that dose would be safe for a human, allowing a safety factor of 10 to allow for interspecies differences between two mammals; if the data are from fish, one might use a factor of 100 to account for the greater difference between two chordate classes (fish and mammals). Similarly, an extra protection factor may be used for individuals believed to be more susceptible to toxic effects such as in pregnancy or with certain diseases. Or, a newly synthesized and previously unstudied chemical that is believed to be very similar in effect to another compound could be assigned an additional protection factor of 10 to account for possible differences in effects that are probably much smaller. Obviously, this approach is very approximate; but such protection factors are deliberately very conservative, and the method has been found to be useful in a deep variety of applications.

Assessing all aspects of the toxicity of cancer-causing agents involves additional issues, since it is not certain if there is a minimal effective dose for carcinogens, or whether the risk is just too small to see. In addition, it is possible that a single cell transformed into a cancer cell is all it takes to develop the full effect (the "one hit" theory).

It is more difficult to determine the toxicity of chemical mixtures than a pure chemical, because each component displays its own toxicity, and components may interact to produce enhanced or diminished effects. Common mixtures include gasoline, cigarette smoke, and industrial waste. Even more complex are situations with more than one type of toxic entity, such as the discharge from a malfunctioning sewage treatment plant, with both chemical and biological agents.

The preclinical toxicity testing on various biological systems reveals the species-, organ- and dose- specific toxic effects of an investigational product. The toxicity of substances can be observed by (a) studying the accidental exposures to a substance (b) in vitro studies using cells/ cell lines (c) in vivo exposure on experimental animals. Toxicity tests are mostly used to examine specific adverse events or specific end points such as cancer, cardiotoxicity, and skin/eye irritation. Toxicity testing also helps calculate the No Observed Adverse Effect Level (NOAEL) dose and is helpful for clinical studies.

Global Classifications of Toxicity

For substances to be regulated and handled appropriately they must be properly classified and labelled. Classification is determined by approved testing measures or calculations and have determined cut off levels set by governments and scientists. While currently many countries have different regulations regarding the types of tests, amounts of tests and cut off levels, the implementation of Global Harmonization will begin unifying these countries as early as 2008.

Global Classification looks at three areas: Physical Hazards (explosions and pyrotechnics), Health Hazards and Environmental Hazards.

Health Hazards

The types of toxicities where substances may cause lethality to the entire body, lethality to specific organs, major/minor damage, or cause cancer. These are globally accepted definitions of what toxicity is. Anything falling outside of the definition cannot be classified as that type of toxicant.

Acute Toxicity

Acute toxicity looks at lethal effects following oral, dermal or inhalation exposure. It is split into five categories of severity where Category 1 requires the least amount of exposure to be lethal and Category 5 requires the most exposure to be lethal. The table below shows the upper limits for each category.

Method of administration	*Category 1*	*Category 2*	*Category 3*	*Category 4*	*Category 5*
Oral: LD_{50} measured in mg/kg of bodyweight	5	50	300	2 000	5 000
Dermal: LD_{50} measured in mg/kg of bodyweight	50	200	1 000	2 000	5 000
Gas Inhalation: LC_{50} measured in ppmV	100	500	2 500	20 000	Undefined
Vapour Inhalation: LC_{50} measured in mg/L	0.5	2.0	10	20	Undefined
Dust and Mist Inhalation: LC_{50} measured in mg/L	0.05	0.5	1.0	5.0	Undefined

Note: The undefined values are expected to be roughly equivalent to the category 5 values for oral and dermal administration.

Other Methods of Exposure and Severity

Skin corrosion and irritation are determined though a skin patch test analysis. This examines the severity of the damage done; when it is incurred and how long it remains; whether it is reversible and how many test subjects were affected.

Skin corrosion from a substance must penetrate through the epidermis into the dermis within four hours of application and must not reverse the damage within 14 days. Skin irritation shows damage less severe than corrosion if: the damage occurs within 72 hours of application; or for three consecutive days after application within a 14 day period; or causes inflammation which lasts for 14 days in two test subjects. Mild skin irritation minor damage (less severe than irritation) within 72 hours of application or for three consecutive days after application.

Serious eye damage involves tissue damage or degradation of vision which does not fully reverse in 21 days. Eye irritation involves changes to the eye which do fully reverse within 21 days.

Other Categories of Toxicity

- Respiratory sensitizers cause breathing hypersensitivity when the substance is inhaled.
- A substance which is a skin sensitizer causes an allergic response from a dermal application.
- Carcinogens induce cancer, or increase the likelihood of cancer occurring.
- Reproductively toxic substances cause adverse effects in either sexual function or fertility to either a parent or the offspring.
- Specific-target organ toxins damage only specific organs.
- Aspiration hazards are solids or liquids which can cause damage through inhalation.

Environmental Hazards

Environmental hazards tend to focus on degradability, bioaccumulation and aquatic toxicity.

Mapping Environmental Hazards

There are many environmental health mapping tools. TOXMAP is a Geographic Information System (GIS) from the Division of Specialised Information Services of the United States National Library of Medicine (NLM) that uses maps of the United States to help users visually explore data from the United States Environmental Protection Agency's (EPA) Toxics Release Inventory and Superfund programs. TOXMAP is a resource funded by the US Federal Government. TOXMAP's chemical and environmental health information is taken from NLM's Toxicology Data Network (TOXNET) and PubMed, and from other authoritative sources.

Aquatic Toxicity

Aquatic toxicity testing submerges key indicator species of fish or crustacea to certain concentrations of a substance in their environment to determine the lethality level. Fish are exposed for 96 hours while crustacea are exposed for 48 hours. While GHS does not define toxicity past 100 mg/l, the EPA currently lists aquatic toxicity as "practically non-toxic" in concentrations greater than 100 ppm.

Exposure	*Category 1*	*Category 2*	*Category 3*
Acute	≤ 1.0 mg/L	≤ 10 mg/L	≤ 100 mg/L
Chronic	≤ 1.0 mg/L	≤ 10 mg/L	≤ 100 mg/L

Note: A category 4 is established for chronic exposure, but simply contains any toxic substance which is mostly insoluble, or has no data for acute toxicity.

Factors Influencing Toxicity

Toxicity of a substance can be affected by many different factors, such as the pathway of administration (whether the toxin is applied to the skin, ingested, inhaled, injected), the time of exposure (a brief encounter or long term), the number of exposures (a single dose or multiple doses over time), the physical form of the toxin (solid, liquid, gas), the genetic makeup of an individual, an individual's overall health, and many others. Several of the terms used to describe these factors have been included here.

Acute exposure a single exposure to a toxic substance which may result in severe biological harm or death; acute exposures are usually characterized as lasting no longer than a day.

Chronic exposure continuous exposure to a toxin over an extended period of time, often measured in months or years; it can cause irreversible side effects.

Chemical Properties of Nanomaterials

Chemical properties of nanomaterials also change at nanoscale. As the percentage of surface atoms in nanoparticles is large compared with bulk objects, therefore reactivities of nanomaterials are more than bulk materials. The following are the some of the chemical properties are:

1.) The preponderance of surface is a major reason for the change in behaviour of materials at the nanoscale. As up to half of all the atoms in nanoparticles are surface atoms, properties such as electrical transport are no longer determined by solid-state bulk phenomenon.

2.) The atoms in nanomaterials have a higher average energy than atoms in longer structures, because of the larger proportion of surface atoms. For example, catalytic materials have a greater chemical activity per atom of exposed surface as the catalyst is reduced in size at the nanoscale.

3.) Defects and impurities may be attracted to surfaces and interfaces, and interactions between particles at those small dimensions can depend on the structure and nature of chemical bonding at the surface.

4.) Molecular monolayers may be used to change or control surface properties and to mediate the interaction between nanoparticles.

Mechanical Properties

The mechanical properties of a material describe how it will react to physical forces. Mechanical properties occur as a result of the physical properties inherent to each material, and are determined through a series of standardized mechanical tests.

Strength

Strength has several definitions depending on the material type and application. Before choosing a material based on its published or measured strength it is important to understand the manner in which strength is defined and how it is measured. When designing for strength, material class and mode of loading are important considerations.

For metals the most common measure of strength is the yield strength. For most polymers it is more convenient to measure the failure strength, the stress at the point where the stress strain curve becomes obviously non-linear. Strength, for ceramics however, is more difficult to define. Failure in ceramics is highly dependent on the mode of loading. The typical failure strength in compression is fifteen times the failure strength in tension. The more common reported value is the compressive failure strength.

Elastic Limit

The elastic limit is the highest stress at which all deformation strains are fully recoverable. For most materials and applications this can be considered the practical limit to the maximum stress a component can withstand and still function as designed. Beyond the elastic limit permanent strains are likely to deform the material to the point where its function is impaired.

Proportional Limit

The proportional limit is the highest stress at which stress is linearly proportional to strain. This is the same as the elastic limit for most materials. Some materials may show a slight deviation from proportionality while still under recoverable strain. In these cases the proportional limit is preferred as a maximum stress level because deformation becomes less predictable above it.

Yield Strength

The yield strength is the minimum stress which produces permanent plastic deformation. This is perhaps the most common material property reported for structural materials because of the ease and relative accuracy of its measurement. The yield strength is usually defined at a specific amount of plastic strain, or offset, which may vary by material and or specification. The offset is the amount that the stress-strain curve deviates from the linear elastic line. The most common offset for structural metals is 0.2%.

Ultimate Tensile Strength

The ultimate tensile strength is an engineering value calculated by dividing the maximum load on a material experienced during a tensile test by the initial cross section of the test sample. When viewed in light of the other tensile test data the ultimate tensile strength helps to provide a good indication of a material's toughness but is not by itself a useful design limit. Conversely this can be construed as the minimum stress that is necessary to ensure the failure of a material.

True Fracture Strength

The true fracture strength is the load at fracture divided by the cross sectional area of the sample. Like the ultimate tensile strength the true fracture strength can help an engineer to predict the behaviour of the material but is not itself a practical strength limit. Because the tensile test seeks to standardize variables such as specimen geometry, strain rate and uniformity of stress it can be considered a kind of best case scenario of failure.

Ductility

Ductility is a measure of how much deformation or strain a material can withstand before breaking. The most common measure of ductility is the percentage of change in length of a tensile sample after breaking. This is generally reported as % El or percent elongation. The R.A. or reduction of area of the sample also gives some indication of ductility.

Toughness

Toughness describes a material's resistance to fracture. It is often expressed in terms of the amount of energy a material can absorb before fracture. Tough materials can absorb a considerable amount of energy before fracture while brittle materials absorb very little. Neither strong materials such as glass or very ductile materials such as taffy can absorb large amounts of energy before failure. Toughness is not a single property but rather a combination of strength and ductility.

The toughness of a material can be related to the total area under its stress-strain curve. A comparison of the relative magnitudes of the yield strength, ultimate tensile strength and percent elongation of different material will give a good indication of their relative toughness. Materials with high yield strength and high ductility have high toughness. Integrated stress-strain data is not readily available for most materials so other test methods have been devised to help quantify toughness. The most common test for toughness is the Charpy impact test.

In crystalline materials the toughness is strongly dependent on crystal structure. Face centred cubic materials are typically ductile while hexagonal close packed materials tend to be brittle. Body centred cubic materials often display dramatic variation in the mode of failure with temperature. In many materials the toughness is temperature dependent. Generally materials are more brittle at lower temperatures and more ductile at higher temperatures. The temperature at which the transition takes place is known as the DBTT, or ductile to brittle transition temperature. The DBTT is measured by performing a series of Charpy impact tests at various temperatures to determine the ranges of brittle and ductile behaviour. Use of alloys below their transition temperature is avoided due to the risk of catastrophic failure.

Fatigue Ratio

The dimensionless fatigue ratio f is the ratio of the stress required to cause failure after a specific number of cycles to the yield stress of a material. Fatigue tests are generally run through 10^7 or 10^8 cycles. A high fatigue ratio indicates materials which are more susceptible to crack growth during cyclic loading.

Loss Coefficient

The loss coefficient is an other important material parameter in cyclic loading. It is the fraction of mechanical energy lost in a stress strain cycle. The loss coefficient for each material is a function of the frequency of the cycle. A high loss coefficient can be desirable for damping vibrations while a low loss coefficient transmits energy more efficiently. The loss coefficient is also an important factor in resisting fatigue failure. If the loss coefficient is too high, cyclic loading will dissipate energy into the material leading to fatigue failure.

2

Thin Film

A thin film is a layer of material ranging from fractions of a nanometre (monolayer) to several micrometres in thickness. Electronic semiconductor devices and optical coatings are the main applications benefiting from thin-film construction. A familiar application of thin films is the household mirror, which typically has a thin metal coating on the back of a sheet of glass to form a reflective interface. The process of silvering was once commonly used to produce mirrors. A very-thin-film coating (less than about 50 nanometres thick) is used to produce two-way mirrors.

The performance of optical coatings (e.g., antireflective, or AR, coatings) are typically enhanced when the thin-film coating consists of multiple layers having varying thicknesses and refractive indices. Similarly, a periodic structure of alternating thin films of different materials may collectively form a so-called superlattice which exploits the phenomenon of quantum confinement by restricting electronic phenomena to two-dimensions.

Work is being done with ferromagnetic and ferroelectric thin films for use as computer memory. It is also being applied to pharmaceuticals, via thin-film drug delivery. Thin-films are used to produce thin-film batteries. Thin films are also used in dye-sensitized solar cells.

Ceramic thin films are in wide use. The relatively high hardness and inertness of ceramic materials make this type of thin coating of interest for protection of substrate materials against corrosion, oxidation and wear. In particular, the use of such coatings on cutting tools can extend the life of these items by several orders of magnitude.

Research is being done on a new class of thin-film inorganic oxide materials, called amorphous heavy-metal cation multicomponent oxides, which could be used to make transparent transistors that are inexpensive, stable, and environmentally benign.

Deposition

The act of applying a thin film to a surface is *thin-film deposition* – any technique for depositing a thin film of material onto a substrate or onto previously deposited layers. "Thin" is a relative term, but most deposition techniques control layer thickness within a few tens of nanometres. Molecular beam epitaxy allows a single layer of atoms to be deposited at a time.

It is useful in the manufacture of optics (for reflective, anti-reflective coatings or self-cleaning glass, for instance), electronics (layers of insulators, semiconductors, and conductors form integrated circuits), packaging (i.e., aluminium-coated PET film), and in contemporary art. Similar processes are sometimes used where thickness is not important: for instance, the purification of copper by electroplating, and the deposition of silicon and enriched uranium by a CVD-like process after gas-phase processing.

Deposition techniques fall into two broad categories, depending on whether the process is primarily chemical or physical.

Chemical Deposition

Here, a fluid precursor undergoes a chemical change at a solid surface, leaving a solid layer. An everyday example is the formation of soot on a cool object when it is placed inside a flame. Since the fluid surrounds the solid object, deposition happens on every surface, with little regard to direction; thin films from chemical deposition techniques tend to be *conformal*, rather than *directional.*

Chemical deposition is further categorized by the phase of the precursor:

- Plating relies on liquid precursors, often a solution of water with a salt of the metal to be deposited. Some plating processes are driven entirely by reagents in the solution (usually for noble metals), but by far the most commercially important process is electroplating. It was not commonly used in semiconductor processing for many years, but has seen a resurgence with more widespread use of chemical-mechanical polishing techniques.
- Chemical solution deposition (CSD) or Chemical bath deposition (CBD) uses a liquid precursor, usually a solution of organometallic powders dissolved in an organic solvent. This is a relatively inexpensive, simple thin film process that is able to produce stoichiometrically accurate crystalline phases. This technique is also known as the sol-gel method

because the 'sol' (or solution) gradually evolves towards the formation of a gel-like diphasic system.

- Spin coating or spin casting, uses a liquid precursor, or sol-gel precursor deposited onto a smooth, flat substrate which is subsequently spun at a high velocity to centrifugally spread the solution over the substrate. The speed at which the solution is spun and the viscosity of the sol determine the ultimate thickness of the deposited film. Repeated depositions can be carried out to increase the thickness of films as desired. Thermal treatment is often carried out in order to crystallize the amorphous spin coated film. Such crystalline films can exhibit certain preferred orientations after crystallization on single crystal substrates.
- Chemical vapor deposition (CVD) generally uses a gas-phase precursor, often a halide or hydride of the element to be deposited. In the case of MOCVD, an organometallic gas is used. Commercial techniques often use very low pressures of precursor gas.
  - o Plasma enhanced CVD (PECVD) uses an ionized vapor, or plasma, as a precursor. Unlike the soot example above, commercial PECVD relies on electromagnetic means (electric current, microwave excitation), rather than a chemical reaction, to produce a plasma.
- Atomic layer deposition (ALD) uses gaseous precursor to deposit conformal thin films one layer at a time. The process is split up into two half reactions, run in sequence and repeated for each layer, in order to ensure total layer saturation before beginning the next layer. Therefore, one reactant is deposited first, and then the second reactant is deposited, during which a chemical reaction occurs on the substrate, forming the desired composition. As a result of the stepwise, the process is slower than CVD, however it can be run at low temperatures, unlike CVD.

Physical Deposition

Physical deposition uses mechanical, electromechanical or thermodynamic means to produce a thin film of solid. An everyday example is the formation of frost. Since most engineering materials are held together by relatively high energies, and chemical reactions are not used to store these energies, commercial physical deposition systems tend to require a low-pressure vapor environment to function properly; most can be classified as physical vapor deposition (PVD).

The material to be deposited is placed in an energetic, entropic environment, so that particles of material escape its surface. Facing this source is a cooler surface which draws energy from these particles as they arrive, allowing them to form a solid layer. The whole system is kept in a vacuum deposition chamber, to allow the particles to travel as freely as possible. Since particles tend to follow a straight path, films deposited by physical means are commonly *directional*, rather than *conformal*.

Examples of physical deposition include:

- A thermal evaporator uses an electric resistance heater to melt the material and raise its vapor pressure to a useful range. This is done in a high vacuum, both to allow the vapor to reach the substrate without reacting with or scattering against other gas-phase atoms in the chamber, and reduce the incorporation of impurities from the residual gas in the vacuum chamber. Obviously, only materials with a much higher vapor pressure than the heating element can be deposited without contamination of the film. Molecular beam epitaxy is a particularly sophisticated form of thermal evaporation.
    - An electron beam evaporator fires a high-energy beam from an electron gun to boil a small spot of material; since the heating is not uniform, lower vapor pressure materials can be deposited. The beam is usually bent through an angle of 270° in order to ensure that the gun filament is not directly exposed to the evaporant flux. Typical deposition rates for electron beam evaporation range from 1 to 10 nanometres per second.
    - In molecular beam epitaxy (MBE), slow streams of an element can be directed at the substrate, so that material deposits one atomic layer at a time. Compounds such as gallium arsenide are usually deposited by repeatedly applying a layer of one element (i.e., gallium), then a layer of the other (i.e., arsenic), so that the process is chemical, as well as physical. The beam of material can be generated by either physical means (that is, by a furnace) or by a chemical reaction (chemical beam epitaxy).
- Sputtering relies on a plasma (usually a noble gas, such as argon) to knock material from a "target" a few atoms at a time. The target can be kept at a relatively low temperature, since the process is not one of evaporation, making this one of the most flexible deposition techniques. It is especially useful for compounds or mixtures, where different components would otherwise tend to evaporate at different rates. Note, sputtering's step coverage is more or less conformal. It is also widely used in the optical media. The manufacturing of all formats of CD, DVD, and BD are done with the help of this technique. It is a fast technique and also it provides a good thickness control. Presently, nitrogen and oxygen gases are also being used in sputtering.

- Pulsed laser deposition systems work by an ablation process. Pulses of focused laser light vaporize the surface of the target material and convert it to plasma; this plasma usually reverts to a gas before it reaches the substrate.
- Cathodic arc deposition (arc-PVD) which is a kind of ion beam deposition where an electrical arc is created that literally blasts ions from the cathode. The arc has an extremely high power density resulting in a high level of ionization (30–100%), multiply charged ions, neutral particles, clusters and macro-particles (droplets). If a reactive gas is introduced during the evaporation process, dissociation, ionization and excitation can occur during interaction with the ion flux and a compound film will be deposited.
- Electrohydrodynamic deposition (electrospray deposition) is a relatively new process of thin film deposition. The liquid to be deposited, either in the form of nano-particle solution or simply a solution, is fed to a small capillary nozzle (usually metallic) which is connected to a high voltage. The substrate on which the film has to be deposited is connected to ground. Through the influence of electric field, the liquid coming out of the nozzle takes a conical shape (Taylor cone) and at the apex of the cone a thin jet emanates which disintegrates into very fine and small positively charged droplets under the influence of Rayleigh charge limit. The droplets keep getting smaller and smaller and ultimately get deposited on the substrate as a uniform thin layer.

Printed Electronics

Printed electronics is a set of printing methods used to create electrical devices on various substrates. Printing typically uses common printing equipment or other low-cost equipment suitable for defining patterns on material, such as screen printing, flexography, gravure, offset lithography, and inkjet. Electrically functional electronic or optical inks are deposited on the substrate, creating active or passive devices, such as thin film transistors or resistors. Printed electronics is expected to facilitate widespread, very low-cost, low-performance electronics for applications such as flexible displays, smart labels, decorative and animated posters, and active clothing that do not require high performance.

The term *printed electronics* is related to organic electronics or plastic electronics, in which one or more inks are composed of carbon-based compounds. These other terms refer to the ink material, which can be deposited by solution-based, vacuum-based or some other method. Printed electronics, in contrast, specifies the process, and can utilise any solution-based material, including organic semiconductors, inorganic semiconductors, metallic conductors, nanoparticles, nanotubes, etc. For the preparation of printed electronics nearly all industrial printing methods are employed. Similar to conventional printing, printed electronics applies ink layers one atop another. so that the coherent development of printing methods and ink materials are the field's essential tasks.

The most important benefit of printing is low-cost volume fabrication. The lower cost enables use in more applications. An example is RFID-systems, which enable contactless identification in trade and transport. In some domains, such as light-emitting diodes printing does not impact performance. Printing on flexible substrates allows electronics to be placed on curved surfaces, for example, putting solar cells on vehicle roofs. More typically, conventional semiconductors justify their much higher costs by providing much higher performance.

Resolution, Registration, Thickness, Holes, Materials

The maximum required resolution of structures in conventional printing is determined by the human eye. Feature sizes smaller than approximately 20 μm cannot be distinguished by the human eye and consequently exceed the capabilities of conventional printing processes. In contrast, higher resolution and smaller structures are necessary in much electronics printing, because they directly affect circuit density and functionality (especially transistors). A similar requirement holds for the precision with which layers are printed on top of each other (layer to layer registration).

Control of thickness, holes, and material compatibility (wetting, adhesion, solvation) are essential, but matter in conventional printing only if the eye can detect them. Conversely, the visual impression is irrelevant for printed electronics.

Printing Technologies

The attraction of printing technology for the fabrication of electronics mainly results from the possibility of preparing stacks of micro-structured layers (and thereby thin-film devices) in a much simpler and cost-effective way compared to conventional electronics. Also, the ability to implement new or improved functionalities (e.g. mechanical flexibility) plays a role. The selection of the printing method used is determined by requirements concerning printed layers, by the properties of printed materials as well as economic and technical considerations of the final printed products.

Printing technologies divide between sheet-based and roll-to-roll-based approaches. Sheet-based inkjet and screen printing are best for low-volume, high-precision work. Gravure, offset and flexographic printing are more common for high-volume production, such as solar cells, reaching 10.000 square metres per hour (m^2/h). While offset and flexographic printing are mainly used for inorganic and organic conductors (the latter also for dielectrics), gravure printing is especially suitable for quality-sensitive layers like organic semiconductors and semiconductor/dielectric-interfaces in transistors, due to high layer quality. In connection with high resolution, is also suitable for inorganic and organic conductors. Organic field-effect transistors and integrated circuits can be prepared completely by means of mass-printing methods.

Inkjets are flexible and versatile, and can be set up with relatively low effort. Inkjets are probably the most commonly used method. However, inkjets offer lower throughput of around 100 m^2/h and lower resolution (ca. 50 μm). It is well suited for low-viscosity, soluble materials like organic semiconductors. With high-viscosity materials, like organic dielectrics, and dispersed particles, like inorganic metal inks, difficulties due to nozzle clogging occur. Because ink is deposited via droplets, thickness and dispersion homogeneity is reduced. Simultaneously using many nozzles and pre-structuring the substrate allows improvements in productivity and resolution, respectively. However, in the latter case non-printing methods must be employed for the actual patterning step. Inkjet printing is preferable for organic semiconductors in organic field-effect transistors (OFETs) and organic light-emitting diodes (OLEDs), but also OFETs completely prepared by this method have been demonstrated. Frontplanes and backplanes of OLED-displays, integrated circuits, organic photovoltaic cells (OPVCs) and other devices can be prepared with inkjets.

Screen printing is appropriate for fabricating electrics and electronics due to its ability to produce patterned, thick layers from paste-like materials. This method can produce conducting lines from inorganic materials (e.g. for circuit boards and antennas), but also insulating and passivating layers, whereby layer thickness is more important than high resolution. Its 50 m^2/h throughput and 100 μm resolution are similar to inkjets. This versatile and comparatively simple method is used mainly for conductive and dielectric layers, but also organic semiconductors, e.g. for OPVCs, and even complete OFETs can be printed.

Aerosol Jet Printing (also known as Maskless Mesoscale Materials Deposition or M3D) is another material deposition technology for printed electronics. The Aerosol Jet process begins with atomization of an ink, which can be heated up to 80 °C, producing droplets on the order of one to two micrometres in diameter. The atomized droplets are entrained in a gas stream and delivered to the print head. Here, an annular flow of clean gas is introduced around the aerosol stream to focus the droplets into a tightly collimated beam of material. The combined gas streams exit the print head through a converging nozzle that compresses the aerosol stream to a diameter as small as 10 μm. The jet of droplets exits the print head at high velocity (~50 metres/second) and impinges upon the substrate. Electrical interconnects, passive and active components are formed by moving the print head, equipped with a mechanical stop/start shutter, relative to the substrate. The resulting patterns can have features ranging from 10 μm wide, with layer thicknesses from tens of nanometres to >10 μm. A wide nozzle print head enables efficient patterning of millimetre size electronic features and surface coating applications. All printing occurs without the use of vacuum or pressure chambers and at room temperature. The high exit velocity of the jet enables a relatively large separation between the print head and the substrate, typically 2–5 mm. The droplets remain tightly focused over this distance, resulting in the ability to print conformal patterns over three dimensional substrates.

Despite the high velocity, the printing process is gentle; substrate damage does not occur and there is generally no splatter or overspray from the droplets. Once patterning is complete, the printed ink typically requires post treatment to attain final electrical and mechanical properties. Post-treatment is driven more by the specific ink and substrate combination than by the printing process. A wide range of materials has been successfully deposited with the Aerosol Jet process, including diluted thick film pastes, thermosetting polymers such as UV-curable epoxies, and solvent-based polymers like polyurethane and polyimide, and biologic materials.

Evaporation printing uses a combination of high precision screen printing with material vaporization to print features to 5 μm. This method uses techniques such as thermal, e-beam, sputter and other traditional production technologies to deposit materials through a high precision shadow mask (or stencil) that is registered to the substrate to better than 1 micrometre. By layering different mask designs and/or adjusting materials, reliable, cost-effective circuits can be built additively, without the use of photolithography, on a production scale that is larger and more efficient than traditional printed electronic technologies.

Other methods with similarities to printing, among them microcontact printing and nano-imprint lithography are of interest. Here, μm- and nm-sized layers, respectively, are prepared by methods similar to stamping with soft and hard forms, respectively. Often the actual structures are prepared subtractively, e.g. by deposition of etch masks or by lift-off processes. For example electrodes for OFETs can be prepared. Sporadically pad printing is used in a similar manner. Occasionally so-called transfer methods, where solid layers are transferred from a carrier to the substrate, are considered printed electronics. Electrophotography is currently not used in printed electronics.

Materials

Both organic and inorganic materials are used for printed electronics. Ink materials must be available in liquid form, for solution, dispersion or suspension. They must function as conductors, semiconductors, dielectrics, or insulators. Material costs must be fit for the application.

Electronic functionality and printability can interfere with each other, mandating careful optimization. For example, a higher molecular weight in polymers enhances conductivity, but diminishes solubility. For printing, viscosity, surface tension and solid content must be tightly controlled. Cross-layer interactions such as wetting, adhesion, and solubility as well as post-deposition drying procedures affect the outcome. Additives often used in conventional printing inks are unavailable, because they often defeat electronic functionality. Material properties largely determine the differences between printed and conventional electronics. Printable materials provide decisive advantages beside printability, such as mechanical flexibility and functional adjustment by chemical modification (e.g. light colour in OLEDs). Printed conductors offer lower conductivity and charge carrier mobility. With a few exceptions, inorganic ink materials are dispersions of metallic or semiconducting micro- and nano-particles. Semiconducting nanoparticles used include silicon and oxide semiconductors. Silicon is also printed as an organic precursor which is then converted by pyrolisis and annealing into crystalline silicon.

PMOS but not CMOS is possible in printed electronics.

Organic Materials

Organic printed electronics integrates knowledge and developments from printing, electronics, chemistry, and materials science, especially from organic and polymer chemistry. Organic materials in part differ from conventional electronics in terms of structure, operation and functionality, which influences device and circuit design and optimization as well as fabrication method.

The discovery of conjugated polymers and their development into soluble materials provided the first organic ink materials. Materials from this class of polymers variously possess conducting, semiconducting, electroluminescent, photovoltaic and other properties. Other polymers are used mostly as insulators and dielectrics.

In most organic materials, hole transport is favoured over electron transport. Recent studies indicate that this is a specific feature of organic semiconductor/dielectric-interfaces, which play a major role in OFETs. Therefore p-type devices should dominate over n-type devices. Durability (resistance to dispersion) and lifetime is less than conventional materials.

Organic semiconductors include the conductive polymers poly(3,4-ethylene dioxitiophene), doped with poly(styrene sulfonate), (PEDOT:PSS) and poly(aniline) (PANI). Both polymers are commercially available in different formulations and have been printed using inkjet, screen and offset printing or screen, flexo and gravure printing, respectively.

Polymer semiconductors are processed using inkjet printing, such as poly(thiopene)s like poly(3-hexylthiophene) (P3HT) and poly(9,9-dioctylfluorene co-bithiophen) (F8T2). The latter material has also been gravure printed. Different electroluminescent polymers are used with inkjet printing, as well as active materials for photovoltaics (e.g. blends of P3HT with fullerene derivatives), which in part also can be deposited using screen printing (e.g. blends of poly(phenylene vinylene) with fullerene derivatives).

Printable organic and inorganic insulators and dielectrics exist, which can be processed with different printing methods.

Inorganic Materials

Inorganic electronics provides highly ordered layers and interfaces that organic and polymer materials cannot provide.

Silver nanoparticles are used with flexo, offset and inkjet. Gold particles are used with inkjet.

A.C. electroluminescent (EL) multi-colour displays can cover many tens of square metres, or be incorporated in watch faces and instrument displays. They involve six to eight printed inorganic layers, including a copper doped phosphor, on a plastic film substrate.

Silicon inks are used commercially in transistor circuits for RFID labels, selective emitters and dopant layers for photovoltaics, and temperature sensors. Fully printed field effect transistors produced at room temperature using silicon nanoparticles have been demonstrated, but the field effect mobility is low compared to crystalline silicon (being similar to that obtained in amorphous silicon transistors). In printed nanoparticle silicon, the low mobility arises from an activated charge transport between nanoparticles, which in turn enables its use as a negative temperature coefficient (NTC) thermistor material. The corresponding varistor characteristics have recently been used to develop a current switching transistor, which works as a two-way switch by changing the conductivity between the different pairs of terminals.

CIGS cells can be printed directly onto molybdenum coated glass sheets.

A printed gallium arsenide germanium solar cell demonstrated 40.7% conversion efficiency, eight times that of the best organic cells, approaching the best performance of crystalline silicon.

Substrates

Printed electronics allows the use of flexible substrates, which lowers production costs and allows fabrication of mechanically flexible circuits. While inkjet and screen printing typically imprint rigid substrates like glass and silicon, mass-printing methods nearly exclusively use flexible foil and paper. Poly(ethylene terephthalate)-foil (PET) is a common choice, due to its low cost and higher temperature stability. Poly(ethylene naphthalate)- (PEN) and poly(imide)-foil (PI) are alternatives. Paper's low costs and manifold applications make it an attractive substrate, however, its high roughness and large absorbency make it problematic for electronics.

Other important substrate criteria are low roughness and suitable wettability, which can be tuned pre-treatment (coating, corona). In contrast to conventional printing, high absorbency is usually disadvantageous.

Applications

Printed electronics are in use or under consideration for:

- Radio frequency identification (RFID) tags
- Monitoring
- Data storage
- Display and visual effects
- Toys

Norwegian company ThinFilm successfully demonstrated roll-to-roll printed organic memory in 2009.

Standards Development and Activities

Technical standards and roadmapping initiatives are intended to facilitate value chain development (for sharing of product specifications, characterization standards, etc.) This strategy of standards development mirrors the approach used by silicon-based electronics over the past 50 years. Initiatives include:

- The IEEE Standards Association has published IEEE 1620-2004 and IEEE 1620.1-2006.
- Similar to the well-established International Technology Roadmap for Semiconductors (ITRS), the International Electronics Manufacturing Initiative (iNEMI) has published a roadmap for printed and other organic electronics.

Magnetic and Electron Transport Properties of Granular Films

Nanomaterials composed of a population of particles dispersed in a matrix represent the building block for the next generation of several technologies: energy storage and conversion, thermal management, electronics, and photovoltaics. When interfaces between particles and matrix are imperfect, the size of the particles may strongly influence the effective linear and nonlinear response of the whole system.

Intensive research is being done on insulating magnetic granular films during the past decade. Spin-dependent tunneling gives rise to important magneto-electrical properties, as tunnel magnetoresistance TMR with high sensitivity to magnetic field and low magnetization coercivity. Therefore, these systems are appealing candidates for next generation high density magnetic memories, low magnetic field sensors, and spin-polarized electrodes in new magnetic tunnel junctions MTJs. Various granular films have been studied with granules of different ferromagnetic FM metals Ni, Co, Fe, CoFe, and NiFe and distinct insulating hosts SiO2, TiO2, ZrO2, Al2O3, and MgO. Between them, especially interesting objects are the discontinuous magnetic layers and multilayers, since two-dimensional 2D geometry favours to more pronounced effects of dipolar coupling between magnetic granules. Experiments on such films show that their magneto-electric properties are mainly controlled by the concentration of magnetic material, or by the nominal thicknesst of a 2D granular layer that is its thickness if it were continuous. Presently, the correlation between the transport processes and the film topology obviously dependent on the nominal thickness of granular layers is the main objective in all these studies. Three principal transitions are observed with varying concentration at fixed and high enough temperatures. The first one is that from metallic to activated electrical conduction, controlled by the structural percolation of metallic granules in the insulating matrix, immediately followed by a second one, that from FM to superferromagnetic SFM state.

The third one is that from SFM to superparamagnetic SPM accompanied by the so-called superspin glass SSG phase at low enough temperatures. The SFM state is attributed to preferable alignment of magnetic moments on randomly distributed separate granules due to the dipolar coupling between them, controlled by the so-called magnetic percolation. In fact, the SFM-SPM transition above a certain glass transition temperature Tg Ref. 8 brings about the most pronounced MR effects, in particular, its sensitivity to low magnetic field. However, up to this moment, certain technical problems prevented more detailed study of granular layers of lower nominal thickness due to their very high resistivity values that could clarify magnetotransport mechanisms in these materials. With this purpose, we study here granular films with low concentration of FM material, below the metal-insulator and SFM-SPM thresholds.

Magnetic and Structural Properties of Nanoparticles

In nanotechnology, a particle is defined as a small object that behaves as a whole unit with respect to its transport and properties. Particles are further classified according to diameter. Coarse particles cover a range between 2,500 and 10,000 nanometres. Fine particles are sized between 100 and 2,500 nanometres. Ultrafine particles, or nanoparticles, are between 1 and 100 nanometres in size. The reason for this double name of the same object is that, during the 1970-80s, when the first thorough fundamental studies with "nanoparticles" were underway in the USA (by Granqvist and Buhrman) and Japan, (within an ERATO Project) they were called "ultrafine particles" (UFP). However, during the 1990s before the National Nanotechnology Initiative was launched in the USA, the new name, "nanoparticle," had become fashionable.

Nanoparticles may or may not exhibit size-related properties that differ significantly from those observed in fine particles or bulk materials. Although the size of most molecules would fit into the above outline, individual molecules are usually not referred to as nanoparticles.

Nanoclusters have at least one dimension between 1 and 10 nanometres and a narrow size distribution. Nanopowders are agglomerates of ultrafine particles, nanoparticles, or nanoclusters. Nanometre-sized single crystals, or single-domain ultrafine particles, are often referred to as nanocrystals. Nanoparticle research is currently an area of intense scientific interest due to a wide variety of potential applications in biomedical, optical and electronic fields.

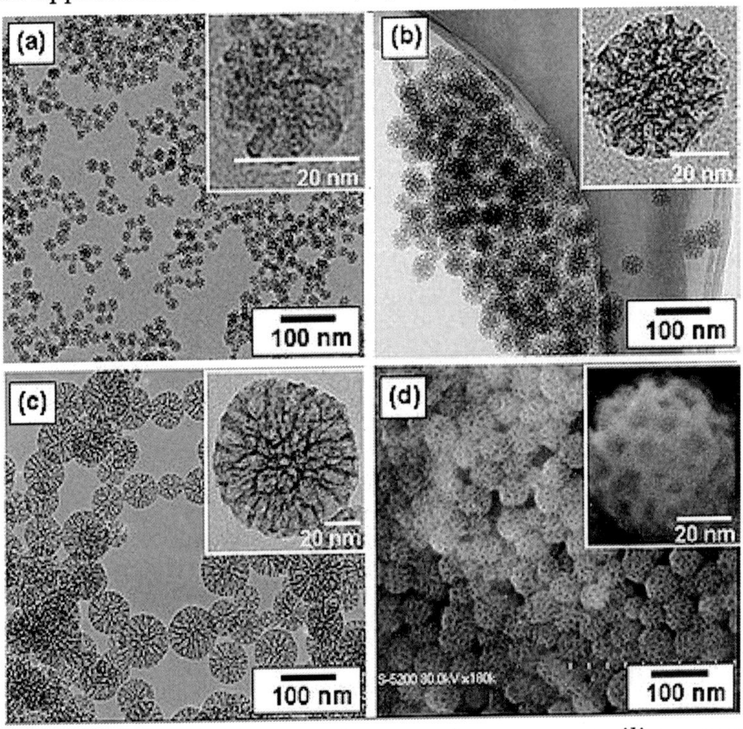

Figure: *TEM (a, b, and c) images of prepared mesoporous silica nanoparticles with mean outer diameter: (a) 20nm, (b) 45nm, and (c) 80nm. SEM (d) image corresponding to (b). The insets are a high magnification of mesoporous silica particle.*

IUPAC Definition

Particle of any shape with dimensions in the 1×10^{-9} and 1×10^{-7} m range.

Note 1: Modified from definitions of nanoparticle and nanogel in.

Note 2: The basis of the 100-nm limit is the fact that novel properties that differentiate particles from the bulk *material* typically develop at a critical length scale of under 100 nm.

Note 3: Because other phenomena (transparency or turbidity, ultrafiltration, stable dispersion, etc.) that extend the upper limit are occasionally considered, the use of the prefix *nano* is accepted for dimensions smaller than 500 nm, provided reference to the definition is indicated.

Note 4: Tubes and fibres with only two dimensions below 100 nm are also nanoparticles.

The National Nanotechnology Initiative has led to generous public funding for nanoparticle research in the United States.

Background

Although, in general, nanoparticles are considered a discovery of modern science, they actually have a very long history. Nanoparticles were used by artisans as far back as the ninth century in Mesopotamia for generating a glittering effect on the surface of pots. Even these days, pottery from the Middle Ages and Renaissance often retain a distinct gold- or copper-coloured metallic glitter. This luster is caused by a metallic film that was applied to the transparent surface of a glazing. The luster can still be visible if the film has resisted atmospheric oxidation and other weathering.

The luster originated within the film itself, which contained silver and copper nanoparticles dispersed homogeneously in the glassy matrix of the ceramic glaze. These nanoparticles were created by the artisans by adding copper and silver salts and oxides together with vinegar, ochre, and clay on the surface of previously-glazed pottery. The object was then placed into a kiln and heated to about 600 °C in a reducing atmosphere. In the heat the glaze would soften, causing the copper and silver ions to migrate into the outer layers of the glaze. There the reducing atmosphere reduced the ions back to metals, which then came together forming the nanoparticles that give the colour and optical effects.

Luster technique showed that ancient craftsmen had a rather sophisticated empirical knowledge of materials. The technique originated in the Muslim world. As Muslims were not allowed to use gold in artistic representations, they sought a way to create a similar effect without using real gold. The solution they found was using luster. Michael Faraday provided the first description, in scientific terms, of the optical properties of nanometre-scale metals in his classic 1857 paper. In a subsequent paper, the author (Turner) points out that: "It is well known that when thin leaves of gold or silver are mounted upon glass and heated to a temperature that is well below a red heat (~500 °C), a remarkable change of properties takes place, whereby the continuity of the metallic film is destroyed. The result is that white light is now freely transmitted, reflection is correspondingly diminished, while the electrical resistivity is enormously increased."

Uniformity

The chemical processing and synthesis of high-performance technological components for the private, industrial, and military sectors requires the use of high-purity ceramics, polymers, glass-ceramics, and composite materials. In condensed bodies formed from fine powders, the irregular particle sizes and shapes in a typical powder often lead to non-uniform packing morphologies that result in packing density variations in the powder compact.

Uncontrolled agglomeration of powders due to attractive van der Waals forces can also give rise to in microstructural inhomogeneities. Differential stresses that develop as a result of non-uniform drying shrinkage are directly related to the rate at which the solvent can be removed, and thus highly dependent upon the distribution of porosity. Such stresses have been associated with a plastic-to-brittle transition in consolidated bodies, and can yield to crack propagation in the unfired body if not relieved.

In addition, any fluctuations in packing density in the compact as it is prepared for the kiln are often amplified during the sintering process, yielding inhomogeneous densification. Some pores and other structural defects associated with density variations have been shown to play a detrimental role in the sintering process by growing and thus limiting end-point densities. Differential stresses arising from inhomogeneous densification have also been shown to result in the propagation of internal cracks, thus becoming the strength-controlling flaws.

Inert gas evaporation and inert gas deposition are free many of these defects due to the distillation (cf. purification) nature of the process and having enough time to form single crystal particles, however even their non-aggreated deposits have lognormal size distribution, which is typical with nanoparticles. The reason why modern gas evaporation techniques can produce a relatively narrow size distribution is that aggregation can be avoided. However, even in this case, random residence times in the growth zone, due to the combination of drift and diffusion, result in a size distribution appearing lognormal.

It would, therefore, appear desirable to process a material in such a way that it is physically uniform with regard to the distribution of components and porosity, rather than using particle size distributions that will maximize the green density. The containment of a uniformly dispersed assembly of strongly interacting particles in suspension requires total control over interparticle forces. Monodisperse nanoparticles and colloids provide this potential.

Monodisperse powders of colloidal silica, for example, may therefore be stabilized sufficiently to ensure a high degree of order in the colloidal crystal or polycrystalline colloidal solid that results from aggregation. The degree of order appears to be limited by the time and space allowed for longer-range correlations to be established. Such defective polycrystalline colloidal structures would appear to be the basic elements of submicrometre colloidal materials science and, therefore, provide the first step in developing a more rigorous understanding of the mechanisms involved in microstructural evolution in high performance materials and components.

Properties

Nanoparticles are of great scientific interest as they are, in effect, a bridge between bulk materials and atomic or molecular structures. A bulk material should have constant physical properties regardless of its size, but at the nano-scale size-dependent properties are often observed. Thus, the properties of materials change as their size approaches the nanoscale and as the percentage of atoms at the surface of a material becomes significant. For bulk materials larger than one micrometre (or micron), the percentage of atoms at the surface is insignificant in relation to the number of atoms in the bulk of the material. *The interesting and sometimes unexpected properties of nanoparticles are therefore largely due to the large surface area of the material, which dominates the contributions made by the small bulk of the material.*

Nanoparticles often possess unexpected optical properties as they are small enough to confine their electrons and produce quantum effects. For example gold nanoparticles appear deep-red to black in solution. Nanoparticles of yellow gold and grey silicon are red in colour. Gold nanoparticles melt at much lower temperatures (~300 °C for 2.5 nm size) than the gold slabs (1064 °C);. Absorption of solar radiation is much higher in materials composed of nanoparticles than it is in thin films of continuous sheets of material. In both solar PV and solar thermal applications, controlling the size, shape, and material of the particles, it is possible to control solar absorption.

Other size-dependent property changes include quantum confinement in semiconductor particles, surface plasmon resonance in some metal particles and superparamagnetism in magnetic materials. What would appear ironic is that the changes in physical properties are not always desirable. Ferromagnetic materials smaller than 10 nm can switch their magnetisation direction using room temperature thermal energy, thus making them unsuitable for memory storage. Suspensions of nanoparticles are possible since the interaction of the particle surface with the solvent is strong enough to overcome density differences, which otherwise usually result in a material either sinking or floating in a liquid.

The high surface area to volume ratio of nanoparticles provides a tremendous driving force for diffusion, especially at elevated temperatures. Sintering can take place at lower temperatures, over shorter time scales than for larger particles. In theory, this does not affect the density of the final product, though flow difficulties and the tendency of nanoparticles to agglomerate complicates matters. Moreover, nanoparticles have been found to impart some extra properties to various day to day products. For example, the presence of titanium dioxide nanoparticles imparts what we call the self-cleaning effect, and, the size being nano-range, the particles cannot be observed. Zinc oxide particles have been found to have superior UV blocking properties compared to its bulk substitute. This is one of the reasons why it is often used in the preparation of sunscreen lotions, and is completely photostable.

Clay nanoparticles when incorporated into polymer matrices increase reinforcement, leading to stronger plastics, verifiable by a higher glass transition temperature and other mechanical property tests. These nanoparticles are hard, and impart their properties to the polymer (plastic). Nanoparticles have also been attached to textile fibres in order to create smart and functional clothing. Metal, dielectric, and semiconductor nanoparticles have been formed, as well as hybrid structures (e.g., core–shell nanoparticles). Nanoparticles made of semiconducting material may also be labelled quantum dots if they are small enough (typically sub 10 nm) that quantization of electronic energy levels occurs. Such nanoscale particles are used in biomedical applications as drug carriers or imaging agents.

Semi-solid and soft nanoparticles have been manufactured. A prototype nanoparticle of semi-solid nature is the liposome. Various types of liposome nanoparticles are currently used clinically as delivery systems for anticancer drugs and vaccines. Nanoparticles with one half hydrophilic and the other half hydrophobic are termed Janus particles and are particularly effective for stabilizing emulsions. They can self-assemble at water/oil interfaces and act as solid surfactants.

Synthesis

There are several methods for creating nanoparticles, including both attrition and pyrolysis. In attrition, macro- or micro-scale particles are ground in a ball mill, a planetary ball mill, or other size-reducing mechanism. The resulting particles are air classified to recover nanoparticles. In pyrolysis, a vaporous precursor (liquid or gas) is forced through an orifice at high pressure and burned. The resulting solid (a version of soot) is air classified to recover oxide particles from by-product gases. Traditional pyrolysis often results in aggregates and agglomerates rather than single primary particles. Ultrasonic nozzle spray pyrolysis (USP) on the other hand aids in preventing agglomerates from forming.

A thermal plasma can also deliver the energy necessary to cause vaporization of small micrometre-size particles. The thermal plasma temperatures are in the order of 10,000 K, so that solid powder easily evaporates. Nanoparticles are formed upon cooling while exiting the plasma region. The main types of the thermal plasma torches used to produce nanoparticles are dc plasma jet, dc arc plasma, and radio frequency (RF) induction plasmas. In the arc plasma reactors, the energy necessary for evaporation and reaction is provided by an electric arc formed between the anode and the cathode. For example, silica sand can be vaporized with an arc plasma at atmospheric pressure. The resulting mixture of plasma gas and silica vapour can be rapidly cooled by quenching with oxygen, thus ensuring the quality of the fumed silica produced.

In RF induction plasma torches, energy coupling to the plasma is accomplished through the electromagnetic field generated by the induction coil. The plasma gas does not come in contact with electrodes, thus eliminating possible sources of contamination and allowing the operation of such plasma torches with a wide range of gases including inert, reducing, oxidizing, and other corrosive atmospheres. The working frequency is typically between 200 kHz and 40 MHz. Laboratory units run at power levels in the order of 30–50 kW, whereas the large-scale industrial units have been tested at power levels up to 1 MW. As the residence time of the injected feed droplets in the plasma is very short, it is important that the droplet sizes are small enough in order to obtain complete evaporation. The RF plasma method has been used to synthesize different nanoparticle materials, for example synthesis of various ceramic nanoparticles such as oxides, carbours/carbides, and nitrides of Ti and Si.

Inert-gas condensation is frequently used to make nanoparticles from metals with low melting points. The metal is vaporized in a vacuum chamber and then supercooled with an inert gas stream. The supercooled metal vapor condenses into nanometre-size particles, which can be entrained in the inert gas stream and deposited on a substrate or studied in situ.

Nanoparticles can also be formed using radiation chemistry. Radiolysis from gamma rays can create strongly active free radicals in solution. This relatively simple technique uses a minimum number of chemicals. These including water, a soluble metallic salt, a radical scavenger (often a secondary alcohol), and a surfactant (organic capping agent). High gamma doses on the order of 10^4 Gray are required. In this process, reducing radicals will drop metallic ions down to the zero-valence state. A scavenger chemical will preferentially interact with oxidizing radicals to prevent the re-oxidation of the metal. Once in the zero-valence state, metal atoms begin to coalesce into particles. A chemical surfactant surrounds the particle during formation and regulates its growth. In sufficient concentrations, the surfactant molecules stay attached to the particle. This prevents it from dissociating or forming clusters with other particles. Formation of nanoparticles using the radiolysis method allows for tailoring of particle size and shape by adjusting precursor concentrations and gamma dose.

Sol-gel

The sol-gel process is a wet-chemical technique (also known as chemical solution deposition) widely used recently in the fields of materials science and ceramic engineering. Such methods are used primarily for the fabrication of materials (typically a metal oxide) starting from a chemical solution (*sol*, short for solution), which acts as the precursor for an integrated network (or *gel*) of either discrete particles or network polymers.

Typical precursors are metal alkoxides and metal chlorides, which undergo hydrolysis and polycondensation reactions to form either a network "elastic solid" or a colloidal suspension (or dispersion) – a system composed of discrete (often amorphous) submicrometre particles dispersed to various degrees in a host fluid. Formation of a metal oxide involves connecting the metal centres with oxo (M-O-M) or hydroxo (M-OH-M) bridges, therefore generating metal-oxo or metal-hydroxo polymers in solution. Thus, the sol evolves toward the formation of a gel-like diphasic system containing both a liquid phase and solid phase whose morphologies range from discrete particles to continuous polymer networks.

In the case of the colloid, the volume fraction of particles (or particle density) may be so low that a significant amount of fluid may need to be removed initially for the gel-like properties to be recognised. This can be accomplished in any number of ways. The most simple method is to allow time for sedimentation to occur, and then pour off the remaining liquid. Centrifugation can also be used to accelerate the process of phase separation.

Removal of the remaining liquid (solvent) phase requires a drying process, which is typically accompanied by a significant amount of shrinkage and densification. The rate at which the solvent can be removed is ultimately determined by the distribution of porosity in the gel. The ultimate microstructure of the final component will clearly be strongly influenced by changes implemented during this phase of processing. Afterward, a thermal treatment, or firing process, is often necessary in order to favour further polycondensation and enhance mechanical properties and structural stability via final sintering, densification, and grain growth. One of the distinct advantages of using this methodology as opposed to the more traditional processing techniques is that densification is often achieved at a much lower temperature.

The precursor sol can be either deposited on a substrate to form a film (e.g., by dip-coating or spin-coating), cast into a suitable container with the desired shape (e.g., to obtain a monolithic ceramics, glasses, fibres, membranes, aerogels), or used to synthesize powders (e.g., microspheres, nanospheres). The sol-gel approach is a cheap and low-temperature technique that allows for the fine control of the product's chemical composition. Even small quantities of dopants, such as organic dyes and rare earth metals, can be introduced in the sol and end up uniformly dispersed in the final product. It can be used in ceramics processing and manufacturing as an investment casting material, or as a means of producing very thin films of metal oxides for various purposes.

Sol-gel derived materials have diverse applications in optics, electronics, energy, space, (bio)sensors, medicine (e.g., controlled drug release) and separation (e.g., chromatography) technology.

Colloids

The term colloid is used primarily to describe a broad range of solid–liquid (and/or liquid–liquid) mixtures, all of which containing distinct solid (and/or liquid) particles that are dispersed to various degrees in a liquid medium. The term is specific to the size of the individual particles, which are larger than atomic dimensions but small enough to exhibit Brownian motion. If the particles are large enough then their dynamic behaviour in any given period of time in suspension would be governed by forces of gravity and sedimentation. But, if they are small enough to be colloids, then their irregular motion in suspension can be attributed to the collective bombardment of a myriad of thermally agitated molecules in the liquid suspending medium, as described originally by Albert Einstein in his dissertation. Einstein proved the existence of water molecules by concluding that this erratic particle behaviour could adequately be described using the theory of Brownian motion, with sedimentation being a possible long-term result. This critical size range (or particle diameter) typically ranges from nanometres (10^{-9} m) to micrometres (10^{-6} m).

Morphology

Scientists have taken to naming their particles after the real-world shapes that they might represent. Nanospheres, nanoreefs, nanoboxes and more have appeared in the literature. These morphologies sometimes arise spontaneously as an effect of a templating or directing agent present in the synthesis such as miscellar emulsions or anodized alumina pores, or from the innate crystallographic growth patterns of the materials themselves. Some of these morphologies may serve a purpose, such as long carbon nanotubes used to bridge an electrical junction, or just a scientific curiosity like the stars shown at right.

Amorphous particles usually adopt a spherical shape (due to their microstructural isotropy), whereas the shape of anisotropic microcrystalline whiskers corresponds to their particular crystal habit. At the small end of the size range, nanoparticles are often referred to as clusters. Spheres, rods, fibres, and cups are just a few of the shapes that have been grown. The study of fine particles is called micromeritics.

Characterization

Nanoparticle characterization is necessary to establish understanding and control of nanoparticle synthesis and applications. Characterization is done by using a variety of different techniques, mainly drawn from materials science. Common techniques are electron microscopy (TEM, SEM), atomic force microscopy (AFM), dynamic light scattering (DLS), x-ray photoelectron spectroscopy (XPS), powder X-ray diffraction (XRD), Fourier transform infrared spectroscopy (FTIR), matrix-assisted laser desorption/ionization time-of-flight mass spectrometry (MALDI-TOF), ultraviolet-visible spectroscopy, dual polarisation interferometry and nuclear magnetic resonance (NMR).

While the theory has been known for over a century, the technology for nanoparticle tracking analysis (NTA) allows direct tracking of the Brownian motion; this method. Therefore, allows the sizing of individual nanoparticles in solution.

Functionalization

The surface coating of nanoparticles is crucial to determining their properties. In particular, the surface coating can regulate stability, solubility, and targeting. A coating that is multivalent or polymeric confers high stability. Functionalized nanomaterial-based catalysts can be used for catalysis of many known organic reactions.

Surface Coating for Biological Applications

For biological applications, the surface coating should be polar to give high aqueous solubility and prevent nanoparticle aggregation. In serum or on the cell surface, highly charged coatings promote non-specific binding, whereas polyethylene glycol linked to terminal hydroxyl or methoxy groups repel non-specific interactions. Nanoparticles can be linked to biological molecules that can act as address tags, to direct the nanoparticles to specific sites within the body, specific organelles within the cell, or to follow specifically the movement of individual protein or RNA molecules in living cells. Common address tags are monoclonal antibodies, aptamers, streptavidin or peptides. These targeting agents should ideally be covalently linked to the nanoparticle and should be present in a controlled number per nanoparticle. Multivalent nanoparticles, bearing multiple targeting groups, can cluster receptors, which can activate cellular signaling pathways, and give stronger anchoring. Monovalent nanoparticles, bearing a single binding site, avoid clustering and so are preferable for tracking the behaviour of individual proteins.

Safety

Nanoparticles present possible dangers, both medically and environmentally. Most of these are due to the high surface to volume ratio, which can make the particles very reactive or catalytic. They are also able to pass through cell membranes in organisms, and their interactions with biological systems are relatively unknown. A recent study looking at the effects of ZnO nanoparticles on human immune cells has found varying levels of susceptibility to cytotoxicity. There are concerns that pharmaceutical companies, seeking regulatory approval for nano-reformulations of existing medicines, are relying on safety data produced during clinical studies of the earlier, pre-reformulation version of the medicine. This could result in regulatory bodies, such as the FDA, missing new side effects that are specific to the nano-reformulation.

Whether cosmetics and sunscreens containing nanomaterials pose health risks remains largely unknown at this stage. However considerable research has demonstrated that zinc nanoparticles are not absorbed into the bloodstream in vivo. Diesel nanoparticles have been found to damage the cardiovascular system in a mouse model.

Concern has also been raised over the health effects of respirable nanoparticles from certain combustion processes. As of 2013 the Environmental Protection Agency was investigating the safety of the following nanoparticles:

- Carbon Nanotubes: Carbon materials have a wide range of uses, ranging from composites for use in vehicles and sports equipment to integrated circuits for electronic components. The interactions between nanomaterials such as carbon nanotubes and natural organic matter strongly influence both their aggregation and deposition, which strongly affects their transport, transformation, and exposure in aquatic environments. In past research, carbon nanotubes exhibited some toxicological impacts that will be evaluated in various environmental settings in current EPA chemical safety research. EPA research will provide data, models, test methods, and best practices to discover the acute health effects of carbon nanotubes and identify methods to predict them.
- Cerium oxide: Nanoscale cerium oxide is used in electronics, biomedical supplies, energy, and fuel additives. Many applications of engineered cerium oxide nanoparticles naturally disperse themselves into the environment, which increases the risk of exposure. There is ongoing exposure to new diesel emissions using fuel additives containing CeO2 nanoparticles, and the environmental and public health impacts of this new technology are unknown. EPA's chemical safety research is assessing the environmental, ecological, and health implications of nanotechnology-enabled diesel fuel additives.
- Titanium dioxide: Nano titanium dioxide is currently used in many products. Depending on the type of particle, it may be found in sunscreens, cosmetics, and paints and coatings. It is also being investigated for use in removing contaminants from drinking water.
- Nano Silver: Nano silver is being incorporated into textiles and other materials to eliminate bacteria and odor from clothing, food packaging, and other items where antimicrobial properties are desirable. In collaboration with the U.S. Consumer Product Safety Commission, EPA is studying certain products to see whether they transfer nano-size silver particles in real-world scenarios. EPA is researching this topic

to better understand how much nano-silver children come in contact with in their environments.

- Iron: While nano-scale iron is being investigated for many uses, including "smart fluids" for uses such as optics polishing and as a better-absorbed iron nutrient supplement, one of its more prominent current uses is to remove contamination from groundwater. This use, supported by EPA research, is being piloted at a number of sites across the country.

Laser Applications

The use of nanoparticle distributions in laser dye-doped poly (methyl methacrylate) (PMMA) laser gain media was demonstrated in 2003 and it has been shown to improve conversion efficiencies and to decrease laser beam divergence. Researchers attribute the reduction in beam divergence to improved dn/dT characteristics of the organic-inorganic dye-doped nanocomposite. The optimum composition reported by these researchers is 30% w/w of SiO_2 (~ 12 nm) in dye-doped PMMA.

Magnetic Nanoparticles

Magnetic nanoparticles are a class of nanoparticle which can be manipulated using magnetic field. Such particles commonly consist of magnetic elements such as iron, nickel and cobalt and their chemical compounds. While nanoparticles are smaller than 1 micrometre in diameter (typically 5–500 nanometres), the larger microbeads are 0.5–500 micrometre in diameter. The magnetic nanoparticles have been the focus of much research recently because they possess attractive properties which could see potential use in catalysis including nanomaterial-based catalysts, biomedicine, magnetic resonance imaging, magnetic particle imaging, data storage, environmental remediation, nanofluids, and optical filters, defect sensor and cation sensors.

Properties

The physical and chemical properties of magnetic nanoparticles largely depend on the synthesis method and chemical structure. In most cases, the particles range from 1 to 100 nm in size and may display superparamagnetism.

Types pf Magnetic Nanoparticles

Oxides: Ferrite: Ferrite nanoparticles are the most explored magnetic nanoparticles up to date. Once the ferrite particles become smaller than 128 nm they become superparamagnetic which prevents self agglomeration since they exhibit their magnetic behaviour only when an external magnetic field is applied. With the external magnetic field switched off, the remanence falls back to zero. Just like non-magnetic oxide nanoparticles, the surface of ferrite nanoparticles is often modified by surfactants, silicones or phosphoric acid derivatives to increase their stability in solution.

Metallic: Metallic nanoparticles have the great disadvantage of being pyrophoric and reactive to oxidizing agents to various degrees. This makes their handling difficult and enables unwanted side reactions.

Metallic with a Shell: The metallic core of magnetic nanoparticles may be passivated by gentle oxidation, surfactants, polymers and precious metals. In an oxygen environment, Co nanoparticles form an anti-ferromagnetic CoO layer on the surface of the Co nanoparticle. Recently, work has explored the synthesis and exchange bias effect in these Co core CoO shell nanoparticles with a gold outer shell. Nanoparticles with a magnetic core consisting either of elementary Iron or Cobalt with a nonreactive shell made of graphene have been synthesized recently. The advantages compared to ferrite or elemental nanoparticles are:

- Higher magnetization
- Higher stability in acidic and basic solution as well as organic solvents
- Chemistry on the graphene surface via methods already known for carbon nanotubes

Synthesis

The established methods of magnetic nanoparticle synthesis include:

Co-Precipitation

Co-precipitation is a facile and convenient way to synthesize iron oxides (either Fe_3O_4 or γ-Fe_2O_3) from aqueous Fe^{2+}/Fe^{3+} salt solutions by the addition of a base under inert atmosphere at room temperature or at elevated temperature. The size, shape, and composition of the magnetic nanoparticles very much depends on the type of salts used (e.g.chlorides, sulfates, nitrates), the Fe^{2+}/Fe^{3+} ratio, the reaction temperature, the pH value and ionic strength of the media, and the mixing rate with the base solution used to provoke the precipitation. The co-precipitation approach has been used extensively to produce ferrite nanoparticles of controlled sizes and magnetic properties. A variety of experimental arrangements have been reported to facilitate continuous and large–scale co–precipitation of magnetic particles by rapid mixing. Recently, the growth rate of the magnetic nanoparticles was measured in real-time during the precipitation of magnetite nanoparticles by an integrated AC magnetic susceptometre within the mixing zone of the reactants.

Thermal Decomposition

Magnetic nanocrystals with smaller size can essentially be synthesized through the thermal decomposition of organometallic compounds in high-boiling organic solvents containing stabilizing surfactants.

Microemulsion

Microemulsions are clear, thermodynamically stable, isotropic liquid mixtures of oil, water and surfactant, frequently in combination with a cosurfactant. The aqueous phase may contain salt(s) and/or other ingredients, and the "oil" may actually be a complex mixture of different hydrocarbons and olefins. In contrast to ordinary emulsions, microemulsions form upon simple mixing of the components and do not require the high shear conditions generally used in the formation of ordinary emulsions. The three basic types of microemulsions are direct (oil dispersed in water, o/w), reversed (water dispersed in oil, w/o) and bicontinuous.

In ternary systems such as microemulsions, where two immiscible phases (water and 'oil') are present with a surfactant, the surfactant molecules may form a monolayer at the interface between the oil and water, with the hydrophobic tails of the surfactant molecules dissolved in the oil phase and the hydrophilic head groups in the aqueous phase.

Uses

Microemulsions have many commercially important uses:

- Water-in-oil microemulsions for some dry cleaning processes
- Floor polishers and cleaners
- Personal care products
- Pesticide formulations
- Cutting oils.
- Drugs.

Much of the work done on these systems have been motivated by their possible use to mobilize petroleum trapped in porous sandstone for enhanced oil recovery A fundamental reason for the uses of these systems is that a microemulsion phase sometimes has an ultralow interfacial tension with a separate oil or aqueous phase, which may release or mobilize them from solid phases even in conditions of slow flow or low pressure gradients.

Microemulsions also have industrial applications, one of them being the synthesis of polymers. Microemulsion polymerization is a complex heterogeneous process where transport of monomers, free radicals and other species (such as chain transfer agent, co-surfactant and inhibitors) between the aqueous and organic phases, takes place. Compared with other heterogeneous polymerization processes (suspension or emulsion) microemulsion polymerization is a more complicated system. Polymerization rate is controlled by monomer partitioning between the phases, particle nucleation, and adsorption and desorption of radicals. Particle stability is affected by the amount and type of surfactant and pH of dispersing medium. It is also used in the process of creating nanoparticles. The kinetics of microemulsion polymerization has much in common with emulsion polymerization kinetics, the most characteristic feature of which is the compartmentalization, where the radicals growing inside the particles are separated from each other, thus suppressing termination to a high extent and, as a consequence, providing high rates of polymerization.

Theory

Various theories concerning microemulsion formation, stability and phase behaviour have been proposed over the years. For example, one explanation for their thermodynamic stability is that the oil/water dispersion is stabilized by the surfactant present and their formation involves the elastic properties of the surfactant film at the oil/water interface, which involves as parameters, the curvature and the rigidity of the film. These parameters may have an assumed or measured pressure and/or temperature dependence (and/or the salinity of the aqueous phase), which may be used to infer the region of stability of the microemulsion, or to delineate the region where three coexisting phases occur, for example. Calculations of the interfacial tension of the microemulsion with a coexisting oil or aqueous phase are also often of special focus and may sometimes be used to guide their formulation.

History and Terminology

The term microemulsion was first used by T. P. Hoar and J. H. Shulman, professors of chemistry at Cambridge University, in 1943. Alternative names for these systems are often used, such as transparent emulsion, swollen micelle, micellar solution, and solubilized oil. More confusingly still, the term microemulsion can refer to the single isotropic phase that is a mixture of oil, water and surfactant, or to one that is in equilibrium with coexisting predominantly oil and/or aqueous phases, or even to other non-isotropic phases. As in the binary systems (water/surfactant or oil/surfactant), self-assembled structures of different types can be formed, ranging, for example, from (inverted) spherical and cylindrical micelles to lamellar phases and bicontinuous microemulsions, which may coexist with predominantly oil or aqueous phases.

Phase Diagrams

Microemulsion domains are usually characterized by constructing ternary-phase diagrams. Three components are the basic requirement to form a microemulsion: two immiscible liquids and a surfactant. The majority of microemulsions use oil and water as immiscible liquid pairs. If a cosurfactant is used, it may sometimes be represented at a fixed ratio to surfactant as a single component, and treated as a single "pseudo-component". The relative amounts of these three components can be represented in a ternary phase diagram. Gibbs phase diagrams can be used to show the influence of changes in the volume fractions of the different phases on the phase behaviour of the system. The three components composing the system are each found at an apex of the triangle, where their corresponding volume fraction is 100%. Moving away from that corner reduces the volume fraction of that specific component and increases the volume fraction of one or both of the two other components. Each point within the triangle represents a possible composition of a mixture of the three components or pseudo-components, which may consist (ideally, according to the Gibbs' phase rule) of one, two or three phases. These points combine to form regions with boundaries between them, which represent the "phase behaviour" of the system at constant temperature and pressure.

The Gibbs phase diagram, however, is an empirical visual observation of the state of the system and may, or may not express the true number of phases within a given composition. Apparently clear single phase formulations can still consist of multiple iso-tropic phases (e.g. the apparently clear heptane/AOT/water microemulsions consist multiple phases). Since these systems can be in equilibrium with other phases, many systems, especially those with high volume fractions of both the two imiscible phases, can be easily destabilised by anything that changes this equilibrium e.g. high or low temperature or addition of surface tension modifying agents.

However, examples of relatively stable microemulsions can be found. It is believed that the mechanism for removing acid build up in car engine oils involves low water phase volume, water-in-oil (w/o) microemulsions. Theoretically, transport of the aqueous acid droplets through the engine oil to microdispersed calcium carbonate particles in the oil should be most efficient when the aqeous droplets are small enough to transport a single hydrogen ion (the smaller the droplets, the greater the number of acid water droplets, the faster the neutralisation). Such microemulsions are probably very stable across a reasonably wide range of elevated temperatures.

Applications

A wide variety of applications have been envisaged for this class of particles which include:

Medical Diagnostics and Treatments

Magnetic nanoparticles are used in an experimental cancer treatment called magnetic hyperthermia in which the fact that nanoparticles heat when they are placed in an alternative magnetic field is used.

Another potential treatment of cancer includes attaching magnetic nanoparticles to free-floating cancer cells, allowing them to be captured and carried out of the body. The treatment has been tested in the laboratory on mice and will be looked at in survival studies.

Magnetic nanoparticles can be used for the detection of cancer. Blood can be inserted onto a microfluidic chip with magnetic nanoparticles in it. These magnetic nanoparticles are trapped inside due to an externally applied magnetic field as the blood is free to flow through. The magnetic nanoparticles are coated with antibodies targeting cancer cells or proteins. The magnetic nanoparticles can be recovered and the attached cancer-associated molecules can be assayed to test for their existence.

Magnetic nanoparticles can be conjugated with carbohydrates and used for detection of bacteria. Iron oxide particles have been used for the detection of Gram negative bacteria like Escherichia coli and for detection of Gram positive bacteria like Streptococcus suis. In an online news story article from Harvard Medical School posted by Jake Miller on Wednesday, March 21, 2012,:

"Researchers from Harvard Medical School and Massachusetts General Hospital have developed a magnetic nanoparticle-based MRI technique for predicting whether—and when—subjects with a genetic predisposition for diabetes will develop the disease. While done initially in mice, preliminary data show that the platform can be used in people as well, so far to distinguish patients that do or do not have pancreas inflammation. "This research is about predicting Type-1 diabetes, and using that predictive power to figure out what is different between those who get it and those who don't get it," said Diane Mathis, Morton Grove-Rasmussen Professor of Immunohematology in the Department of Microbiology and Immunobiology and, along with Christophe Benoist, Morton Grove-Rasmussen Professor of Immunohematology, co-senior author of the paper. The results were published online in Nature Immunology on Feb. 26, 2012. According to first author Wenxian Fu, a research fellow in the Mathis-Benoist lab, the group was surprised that the diagnostic window—from six to 10 weeks of age— was so early, and so brief. This shows that the progression of the disease, at least in this animal model, is determined very early in life, and that diabetes does not require an additional trigger such as a secondary infection or environmental stress ..."

Magnetic Immunoassay

Magnetic immunoassay (MIA) is a novel type of diagnostic immunoassay utilising magnetic beads as labels in lieu of conventional, enzymes, radioisotopes or fluorescent moieties. This assay involves the specific binding of an antibody to its antigen, where a magnetic label is conjugated to one element of the pair. The presence of magnetic beads is then detected by a magnetic reader (magnetometre) which measures the magnetic field change induced by the beads. The signal measured by the magnetometre is proportional to the analyte (virus, toxin, bacteria, cardiac marker,etc.) quantity in the initial sample.

Waste Water Treatment

Thanks to the easy separation by applying a magnetic field and the very large surface to volume ratio, magnetic nanoparticles have a good potential for treatment of contaminated water. In this method, attachment of EDTA-like chelators to carbon coated metal nanomagnets results in a magnetic reagent for the rapid removal of heavy metals from solutions or contaminated water by three orders of magnitude to concentrations as low as micrograms per Litre.

Chemistry

Magnetic nanoparticles are being used or have the potential use as a catalyst or catalyst supports. In chemistry, a catalyst support is the material, usually a solid with a high surface area, to which a catalyst is affixed. The reactivity of heterogeneous catalysts occurs at the surface atoms. Consequently great effort is made to maximize the surface area of a catalyst by distributing it over the support. The support may be inert or participate in the catalytic reactions. Typical supports include various kinds of carbon, alumina, and silica.

Biomedical Imaging

There are many applications for iron-oxide based nanoparticles in concert with magnetic resonance imaging. Magnetic CoPt nanoparticles are being used as an MRI contrast agent for transplanted neural stem cell detection.

Information Storage

Research is going into the use of using MNPs for magnetic recording media. The most promising candidate for high-density storage is the face-centred tetragonal phase FePt alloy. Grain sizes can be as small as 3 nanometres. If it's possible to modify the MNPs at this small scale, the information density that can be achieved with this media could easily surpass 1 Terabyte per square inch.

Genetic Engineering

Magnetic nanoparticles can be used for a variety of genetics applications. One application is the isolation of mRNA. This can be done quickly – usually within 15 minutes. In this particular application, the magnetic bead is attached to a poly T tail. When mixed with mRNA, the poly A tail of the mRNA will attach to the bead's poly T tail and the isolation takes place simply by placing a magnet on the side of the tube and pouring out the liquid. Magnetic beads have also been used in plasmid assembly. Rapid genetic circuit construction has been achieved by the sequential addition of genes onto a growing genetic chain, using nanobeads as an anchor. This method has been shown to be much faster than previous methods, taking less than an hour to create functional multi-gene constructs in vitro.

Physical and Chemical Properties of Nanoparticles

The principal parameters of nanoparticles are their shape (including aspect ratios where appropriate), size, and the morphological sub-structure of the substance. Nanoparticles are presented as an aerosol (mostly solid or liquid phase in air), a suspension (mostly solid in liquids) or an emulsion (two liquid phases). In the presence of chemical agents (surfactants), the surface and interfacial properties may be modified. Indirectly such agents can stabilise against coagulation or aggregation by conserving particle charge and by modifying the outmost layer of the particle. Depending on the growth history and the lifetime of a nanoparticle, very complex compositions, possibly with complex mixtures of adsorbates, have to be expected. In the typical history of a combustion nanoparticle, for example, many different agents are prone to condensation on the particle while it cools down and is exposed to different ambient atmospheres. Complex surface chemical processes are to be expected and have been identified only for a small number of particulate model systems. At the nanoparticle - liquid interface, polyelectrolytes have been utilised to modify surface properties and the interactions between particles and their environment. They have been used in a wide range of technologies, including adhesion, lubrication, stabilization, and controlled flocculation of colloidal dispersions.

At some point between the Angstrom level and the micrometre scale, the simple picture of a nanoparticle as a ball or droplet changes. Both physical and chemical properties are derived from atomic and molecular origin in a complex way. For example the electronic and optical properties and the chemical reactivity of small clusters are completely different from the better known property of each component in the bulk or at extended surfaces. Complex quantum mechanical models are required to predict the evolution of such properties with particle size, and typically very well defined conditions are needed to compare experiments and theoretical predictions.

At the nanoscale, particle-particle interactions are either dominated by weak Van der Waals forces, stronger polar and electrostatic interactions or covalent interactions. Depending on the viscosity and polarisability of the fluid, particle aggregation is determined by the interparticle interaction. By the modification of the surface layer, the tendency of a colloid to coagulate can be enhanced or hindered. For nanoparticles suspended in air, charges can be accumulated by physical processes such as glow discharge or photoemission. In liquids, particle charge can be stabilised by electrochemical processes at surfaces. The details of nanoparticle - nanoparticle interaction forces and nanoparticle – fluid interactions are of key importance to describe physical and chemical processes, and the temporal evolution of free nanoparticles. They remain difficult to characterise due to the small amount of molecules involved in the surface active layer. Both surface energy, charge and solvation are relevant parameters to be considered. Due to the crucial role of the nanoparticle – nanoparticle interaction and the nanoparticle – fluid interaction, the term free nanoparticle can be easily misunderstood. The interaction forces, either attractive or repulsive, crucially determine the fate of individual and collective nanoparticles. This interaction between nanoparticles resulting in aggregates and/or agglomerates may influence on their behaviour. In gas suspensions, aggregation is crucially determined by the size and diffusion, and coagulation typically occurs faster than in the liquid phase as the sticking coefficient is closer to unity than in liquids.

3

Consolidation of Nanocrystalline Materials by Compaction

A nanocrystalline (NC) material is a polycrystalline material with a crystallite size of only a few nanometres. These materials fill the gap between amorphous materials without any long range order and conventional course-grained materials. While there is no universally agreed upon definition, a common definition in the literature is a crystallite (grain) size below 100 nm. Grain sizes from 100–500 nm are typically considered "ultrafine" grains.

The grain size of a NC sample can be estimated using x-ray diffraction. In materials with very small grain sizes, the diffraction peaks will be broadened. This broadening can be related to a crystallite size using the Scherrer equation (applicable up to ~50 nm), a Williamson-Hall plot, or more sophisticated methods such as the Warren-Averbach method or computer modelling of the diffraction pattern. The crystallite size can be measured directly using transmission electron microscopy.

Ultrahigh Strength Ductile Nanocrystalline Steel

Various unconventional properties such as ultrahigh strength, high hardness, superplasticity and unique magnetic properties were reported from nanocrystalline materials whose grain size is less than 100 nm. During the past two decades, many techniques have been developed to synthesize nanocrystalline materials, including inert gas condensation and consolidation (IGC), severe plastic deformation (SPD), crystallization of amorphous solid, electrodeposition, and surface mechanical attrition treatment (SMAT). However, producing industrially useful bulk materials with the microstructure in a nanometre scale is still a challenge for materials engineering. One of the widely-used promising methods to produce commercial bulk nanomaterials with a grain size of less than 100 nm is ball milling accompanied by a subsequent consolidation process. Due to the metastable nature of the nanocrystalline materials, the consolidation process must be able to eliminate extensive grain growth and retain the nanoscale grain size while obtaining a near theoretical density. Numerous traditional consolidation techniques, such as hot pressing, hot isostatic pressing and hot-extrusion can yield nearly fully dense compacts; however, recrystallization and grain growth during high temperature consolidation processes retard the synthesis of nanocrystalline microstructure. An artifact-free bulk nanocrystalline Cu with a good combination of strength (770 MPa) and ductility (30%) was developed very recently by mechanical milling and in-situ consolidation; however, the uncontrollable sample shape and size may be the restriction for engineering applications. A process that can be applied to produce bulk nanocrystalline materials at moderate temperature with low applied load and fast consolidation speed are necessary for potential engineering applications. Spark plasma sintering (SPS) can achieve rapid heating rate with pulsed electrical discharge that is required to densify powders rapidly without substantial grain growth. The SPS method has been used previously to consolidate a large variety of ceramic and metal powder materials to high densities. The SPS method was used very recently to synthesize nanocrystalline Fe-Al with a near theoretical density with controlled grain growth. The unique boundary cleaning effect of the SPS process makes the powder handling more convenient. In the present investigation, we selected Fe-C powder as the precursor to fabricate dense bulk nanocrystalline materials by means of the SPS method, because a recent study showed that the Fe-C nanocrystalline microstructure that was produced by mechanical milling exhibits strong resistance to the grain growth by heating. In addition, the Fe-C system has been used in wide industrial applications and is not susceptible to media pollution during mechanical milling with stainless steel balls.

Fe-0.8wt%C alloy powder was prepared by ball milling iron (99.99% purity) and graphite powder (99.999% purity) in a Fritsch P-6 planetary ball mill. Stainless steel media (10 mm in diameter) were used with a ball to a charge ratio of 10:1. Powder handling was conducted in an argon glove box to avoid contamination. Milling was conducted at 250 rpm for 100 h, with an intermediate stopping for 2 h after every 20 h milling to reduce temperature rise. The as-milled Fe-C powder was packed in a graphite die with an inner diameter of 10 mm and was densified by a SPS apparatus, Sumitomo Coal Mining Company Model 1050, in a vacuum of <10-3 Pa at a load of 5.5 kN (70 MPa) for 10 min. Temperatures of 400oC, 500oC, and 600oC were selected to compare the consolidation behaviour, mechanical properties, and constituent phases. A thermocouple placed in the middle part of the graphite die was used to control the temperature during the SPS treatment.

Densification During Pressureless Sintering

Sintering

Sintering is a method for creating objects from powders, including metal and ceramic powders. It is based on atomic diffusion. Diffusion occurs in any material above absolute zero, but it occurs much faster at higher temperatures. In most sintering processes, the powdered material is held in a mold and then heated to a temperature below the melting point. The atoms in the powder particles diffuse across the boundaries of the particles, fusing the particles together and creating one solid piece. Because the sintering temperature does not have to reach the melting point of the material, sintering is often chosen as the shaping process for materials with extremely high melting points such as tungsten and molybdenum.

Sintering is traditionally used for manufacturing ceramic objects but finds applications in almost all fields of industry. The study of sintering and of powder-related processes is known as powder metallurgy. A simple, intuitive example of sintering can be observed when ice cubes in a glass of water adhere to each other.

Advantages

Particular advantages of the powder technology include:

1. Very high levels of purity and uniformity in starting materials
2. Preservation of purity, due to the simpler subsequent fabrication process (fewer steps) that it makes possible
3. Stabilization of the details of repetitive operations, by control of grain size during the input stages
4. Absence of binding contact between segregated powder particles – or "inclusions" (called stringering) – as often occurs in melting processes
5. No deformation needed to produce directional elongation of grains
6. Capability to produce materials of controlled, uniform porosity.
7. Capability to produce nearly net-shaped objects.
8. Capability to produce materials which cannot be produced by any other technology.
9. Capability to fabricate high-strength material like turbine blades.
10. After sintering the mechanical strength to handling becomes higher.

The literature contains many references on sintering dissimilar materials to produce solid/solid-phase compounds or solid/melt mixtures at the processing stage. Almost any substance can be obtained in powder form, through either chemical, mechanical or physical processes, so basically any material can be obtained through sintering. When pure elements are sintered, the leftover powder is still pure, so it can be recycled.

Disadvantages

Particular disadvantages of the powder technology include:

1. 100% sintered (iron ore) can not be charged in the blast furnace.
2. By sintering one cannot create uniform sizes.

General Sintering

Sintering is effective when the process reduces the porosity and enhances properties such as strength, electrical conductivity, translucency and thermal conductivity; yet, in other cases, it may be useful to increase its strength but keep its gas absorbency constant as in filters or catalysts. During the firing process, atomic diffusion drives powder surface elimination in different stages, starting from the formation of necks between powders to final elimination of small pores at the end of the process.

The driving force for densification is the change in free energy from the decrease in surface area and lowering of the surface free energy by the replacement of solid-vapor interfaces. It forms new but lower-energy solid-solid interfaces with a total decrease in free energy occurring on sintering 1-micrometre particles a 1 cal/g decrease. On a microscopic scale, material transfer is affected by the change in pressure and differences in free energy across the curved surface. If the size of the particle is small (and its curvature is high), these effects become very large in magnitude. The change in energy is much higher when the radius of curvature is less than a few micrometres, which is one of the main reasons why much ceramic technology is based on the use of fine-particle materials.

For properties such as strength and conductivity, the bond area in relation to the particle size is the determining factor. The variables that can be controlled for any given material are the temperature and the initial grain size, because the vapor pressure depends upon temperature. Through time, the particle radius and the vapor pressure are proportional to $(p_0)^{1/2}$ and to $(p_0)^{1/3}$, respectively.

The source of power for solid-state processes is the change in free or chemical potential energy between the neck and the surface of the particle. This energy creates a transfer of material through the fastest means possible; if transfer were to take place from the particle volume or the grain boundary between particles, then there would be particle reduction and pore destruction. The pore elimination occurs faster for a trial with many pores of uniform size and higher porosity where the boundary diffusion distance is smaller. For the latter portions of the process, boundary and lattice diffusion from the boundary become important.

Control of temperature is very important to the sintering process, since grain-boundary diffusion and volume diffusion rely heavily upon temperature, the size and distribution of particles of the material, the materials composition, and often the sintering environment to be controlled.

Ceramic Sintering

Sintering is part of the firing process used in the manufacture of pottery and other ceramic objects. These objects are made from substances such as glass, alumina, zirconia, silica, magnesia, lime, beryllium oxide and ferric oxide. Some ceramic raw materials have a lower affinity for water and a lower plasticity index than clay, requiring organic additives in the stages before sintering. The general procedure of creating ceramic objects via sintering of powders includes:

- Mixing water, binder, deflocculant, and unfired ceramic powder to form a slurry;
- Spray-drying the slurry;
- Putting the spray dried powder into a mold and pressing it to form a *green body* (an unsintered ceramic item);
- Heating the green body at low temperature to burn off the binder;
- Sintering at a high temperature to fuse the ceramic particles together.

All the characteristic temperatures associated to phases transformation, glass transitions and melting points, occurring during a sinterisation cycle of a particular ceramics formulation (i.e., tails and frits) can be easily obtained by observing the expansion-temperature curves during optical dilatometer thermal analysis. In fact, sinterisation is associated to a remarkable shrinkage of the material because glass phases flow, once their transition temperature is reached, and start consolidating the powdery structure and considerably reducing the porosity of the material.

There are two types of sintering: with pressure (also known as hot pressing), and without pressure. Pressureless sintering is possible with graded metal-ceramic composites, with a nanoparticle sintering aid and bulk molding technology. A variant used for 3D shapes is called hot isostatic pressing.

To allow efficient stacking of product in the furnace during sintering and prevent parts sticking together, many manufacturers separate ware using Ceramic Powder Separator Sheets. These sheets are available in various materials such as alumina, zirconia and magnesia. They are additionally categorized by fine, medium and coarse particle sizes. By matching the material and particle size to the ware being sintered, surface damage and contamination can be reduced while maximizing furnace loading.

Sintering of Metallic Powders

Most, if not all, metals can be sintered. This applies especially to pure metals produced in vacuum which suffer no surface contamination. Sintering under atmospheric pressure requires the use of a protective gas, quite often endothermic gas. Sintering, with subsequent reworking, can produce a great range of material properties. Changes in density, alloying, or heat treatments can alter the physical characteristics of various products. For instance, the Young's Modulus E_n of sintered iron powders remains insensitive to sintering time, alloying, or particle size in the original powder, but depends upon the density of the final product:

$$E_n / E = (D / d)^{3.4}$$

where D is the density, E is Young's modulus and d is the maximum density of iron.

Sintering is static when a metal powder under certain external conditions may exhibit coalescence, and yet reverts to its normal behaviour when such conditions are removed. In most cases, the density of a collection of grains increases as material flows into voids, causing a decrease in overall volume. Mass movements that occur during sintering consist of the reduction of total porosity by repacking, followed by material transport due to evaporation and condensation from diffusion. In the final stages, metal atoms move along crystal boundaries to the walls of internal pores, redistributing mass from the internal bulk of the object and smoothing pore walls. Surface tension is the driving force for this movement.

A special form of sintering, still considered part of powder metallurgy, is liquid-state sintering. In liquid-state sintering, at least one but not all elements are in a liquid state. Liquid-state sintering is required for making cemented carbide or tungsten carbide.

Sintered bronze in particular is frequently used as a material for bearings, since its porosity allows lubricants to flow through it or remain captured within it. For materials that have high melting points such as molybdenum, tungsten, rhenium, tantalum, osmium and carbon, sintering is one of the few viable manufacturing processes. In these cases, very low porosity is desirable and can often be achieved.

Sintered metal powder is used to make frangible shotgun shells called breaching rounds, as used by military and SWAT teams to quickly force entry into a locked room. These shotgun shells are designed to destroy door deadbolts, locks and hinges without risking lives by ricocheting or by flying on at lethal speed through the door. They work by destroying the object they hit and then dispersing into a relatively harmless powder.

Sintered bronze and stainless steel are used as filter materials in applications requiring high temperature resistance while retaining the ability to regenerate the filter element. For example, sintered stainless steel elements are employed for filtering steam in food and pharmaceutical applications, and sintered bronze in aircraft hydraulic systems.

Sintering of powders containing precious metals such as silver and gold is used to make small jewellery items.

Plastics Sintering

Plastic materials are formed by sintering for applications that require materials of specific porosity. Sintered plastic porous components are used in filtration and to control fluid and gas flows. Sintered plastics are used in applications requiring wicking properties, such as marking pen nibs. Sintered ultra high molecular weight polyethylene materials are used as ski and snowboard base materials. The porous texture allows wax to be retained within the structure of the base material, thus providing a more durable wax coating.

Liquid Phase Sintering

For materials which are hard to sinter a process called liquid phase sintering is commonly used. Materials for which liquid phase sintering is common are Si_3N_4, WC, SiC, and more. Liquid phase sintering is the process of adding an additive to the powder which will melt before the matrix phase. The process of liquid phase sintering has three stages:

- Rearrangement – As the liquid melts capillary action will pull the liquid into pores and also cause grains to rearrange into a more favourable packing arrangement.
- Solution-Precipitation – In areas where capillary pressures are high (particles are close together) atoms will preferentially go into solution and then precipitate in areas of lower chemical potential where particles are non close or in contact. This is called "*contact flattening*" This densifies the system in a way similar to grain boundary diffusion in solid state sintering. Ostwald ripening will also occur where smaller particles will go into solution preferentially and precipitate on larger particles leading to densification.
- Final Densification – densification of solid skeletal network, liquid movement from efficiently packed regions into pores.

For liquid phase sintering to be practical the major phase should be at least slightly soluble in the liquid phase and the additive should melt before any major sintering of the solid particulate network occurs, otherwise rearrangement of grains will not occur.

Electric Current Assisted Sintering

These techniques employ electric currents to drive or enhance sintering. English engineer A. G. Bloxam registered in 1906 the first patent on sintering powders using direct current in vacuum. The primary purpose of his inventions was the industrial scale production of filaments for incandescent lamps by compacting tungsten or molybdenum particles. The applied current was particularly effective in reducing surface oxides that increased the emissivity of the filaments.

In 1913, Weintraub and Rush patented a modified sintering method which combined electric current with pressure. The benefits of this method were proved for the sintering of refractory metals as well as conductive carbide or nitride powders. The starting boron–carbon or silicon–carbon powders were placed in an electrically insulating tube and compressed by two rods which also served as electrodes for the current. The estimated sintering temperature was 2000 °C.

In the US, sintering was first patented by Duval d'Adrian in 1922. His three-step process aimed at producing heat-resistant blocks from such oxide materials as zirconia, thoria or tantalia. The steps were: (i) molding the powder; (ii) annealing it at about 2500 °C to make it conducting; (iii) applying current-pressure sintering as in the method by Weintraub and Rush.

Sintering which uses an arc produced via a capacitance discharge to eliminate oxides before direct current heating, was patented by G. F. Taylor in 1932. This originated sintering methods employing pulsed or alternating current, eventually superimposed to a direct current. Those techniques have been developed over many decades and summarized in more than 640 patents.

Of these technologies the most well known is resistance sintering (also called hot pressing) and spark plasma sintering, while capacitor discharge sintering is the latest advancement in this field.

Spark Plasma Sintering

In spark plasma sintering (SPS), external pressure and an electric field are applied simultaneously to enhance the densification of the metallic/ceramic powder compacts. This densification uses lower temperatures and shorter amount of time than typical sintering. For a number of years, it was speculated that the existence of sparks or plasma between particles could aid sintering; however, Hulbert and coworkers systematically proved that the electric parameters used during spark plasma sintering make it (highly) unlikely. In light of this, the name "spark plasma sintering" has been rendered obsolete. Terms such as "Field Assisted Sintering Technique" (FAST), "Electric Field

Assisted Sintering" (EFAS), and Direct Current Sintering (DCS) have been implemented by the sintering community. Using a DC pulse as the electrical current, spark plasma, spark impact pressure, joule heating, and an electrical field diffusion effect would be created.

Certain ceramic materials have low density, chemical inertness, high strength, hardness and temperature capability; nanocrystalline ceramics have even greater strength and higher superplasticity.

Many microcrystalline ceramics that were treated and had gained facture toughness lost their strength and hardness, with this many have created ceramic composites to offset the deterioration while increasing strength and hardness to that of nanocrystalline materials. [???] Through various experiments it has been found that in order to design the mechanical properties of new material, controlling the grain size and its distribution, amount of distribution and other is critical.

Pressureless Sintering

Pressureless sintering is the sintering of a powder compact (sometimes at very high temperatures, depending on the powder) without applied pressure. This avoids density variations in the final component, which occurs with more traditional hot pressing methods.

The powder compact (if a ceramic) can be created by slip casting into a plaster mould, then the final green compact can be machined if necessary to final shape before being heated to sinter.

Densification, Vitrification and Grain Growth

Sintering in practice is the control of both densification and grain growth. Densification is the act of reducing porosity in a sample thereby making it more dense. Grain growth is the process of grain boundary motion and Ostwald ripening to increase the average grain size. Many properties (mechanical strength, electrical breakdown strength, etc.) benefit from both a high relative density and a small grain size. Therefore, being able to control these properties during processing is of high technical importance. Since densification of powders requires high temperatures, grain growth naturally occurs during sintering. Reduction of this process is key for many engineering ceramics.

For densification to occur at a quick pace it is essential to have (1) an amount of liquid phase that is large in size, (2) a near complete solubility of the solid in the liquid, and (3) wetting of the solid by the liquid. The power behind the densification is derived from the capillary pressure of the liquid phase located between the fine solid particles. When the liquid phase wets the solid particles, each space between the particles becomes a capillary in which a substantial capillary pressure is developed. For submicrometre particle sizes, capillaries with diameters in the range of 0.1 to 1 micrometres develop pressures in the range of 175 pounds per square inch (1,210 kPa) to 1,750 pounds per square inch (12,100 kPa) for silicate liquids and in the range of 975 pounds per square inch (6,720 kPa) to 9,750 pounds per square inch (67,200 kPa) for a metal such as liquid cobalt.

Densification requires constant capillary pressure where just solution-precipitation material transfer would not produce densification. For further densification, additional particle movement while the particle undergoes grain-growth and grain-shape changes occurs. Shrinkage would result when the liquid slips between particles and increase pressure at points of contact causing the material to move away from the contact areas forcing particle centres to draw near each other.

The sintering of liquid-phase materials involve a fine-grained solid phase to create the needed capillary pressures proportional to its diameter and the liquid concentration must also create the required capillary pressure within range, else the process ceases. The vitrification rate is dependent upon the pore size, the viscosity and amount of liquid phase present leading to the viscosity of the overall composition, and the surface tension. Temperature dependence for densification controls the process because at higher temperatures viscosity decreases and increases liquid content. Therefore, when changes to the composition and processing are made, it will affect the vitrification process.

Sintering Mechanisms

Sintering occurs by diffusion of atoms through the microstructure. This diffusion is caused by a gradient of chemical potential – atoms move from an area of higher chemical potential to an area of lower chemical potential. The different paths the atoms take to get from one spot to another are the sintering mechanisms. The six common mechanisms are:

- Surface diffusion – Diffusion of atoms along the surface of a particle
- Vapor transport – Evaporation of atoms which condense on a different surface
- Lattice diffusion from surface – atoms from surface diffuse through lattice
- Lattice diffusion from grain boundary – atom from grain boundary diffuses through lattice
- Grain boundary diffusion – atoms diffuse along grain boundary
- Plastic deformation – dislocation motion causes flow of matter

Also one must distinguish between densifying and non-densifying mechanisms. 1–3 above are non-densifying – they take atoms from the surface and rearrange them onto another surface or part of the same surface. These mechanisms simply rearrange matter inside of porosity and do not cause pores to shrink. Mechanisms 4–6 are densifying mechanisms – atoms are moved from the bulk to the surface of pores thereby eliminating porosity and increasing the density of the sample.

Grain Growth

A grain boundary(GB) is the transition area or interface between adjacent crystallites (or grains) of the same chemical and lattice composition, not to be confused with a phase boundary. The adjacent grains do not have the same orientation of the lattice thus giving the atoms in GB shifted positions relative to the lattice in the crystals. Due to the shifted positioning of the atoms in the GB they have a higher energy state when compared with the atoms in the crystal lattice of the grains. It is this imperfection that makes it possible to selectively etch the GBs when one wants the microstructure visible. Striving to minimize its energy leads to the coarsening of the microstructure to reach a metastable state within the specimen. This involves minimizing its GB area and changing its topological structure to minimize its energy. This grain growth can either be normal or abnormal, a normal grain growth is characterized by the uniform growth and size of all the grains in the specimen. Abnormal growth is when a few grains grow much larger than the remaining majority.

Grain Boundary Energy/Tension

The atoms in the GB are normally in a higher energy state than their equivalent in the bulk material. This is due to their more stretched bonds, which gives rise to a GB tension σ_{GB}. This extra energy that the atoms possess is called the grain boundary energy, γ_{GB}. The grain will want to minimize this extra energy thus striving to make the grain boundary area smaller and this change requires energy.

"Or, in other words, a forc $E_n / E = (D / d)^{3.4}$ e has to be applied, in the plane of the grain boundary and acting along a line in the grain-boundary area, in order to extend the grain-boundary area in the direction of the force. The force per unit length, i.e. tension/stress, along the line mentioned is σGB. On the basis of this reasoning it would follow:

$$\sigma_{GB} dA \text{ (work done)} = \gamma_{GB} dA \text{ (energy change)}$$

with dA as the increase of grain-boundary area per unit length along the line in the grain-boundary area considered." [pg 478]

The GB tension can also be thought of as the attractive forces between the atoms at the surface and the tension between these atoms is due to the fact that there is a larger interatomic distance between them at the surface compared to the bulk (i.e. surface tension). When the surface area becomes bigger the bonds stretch more and the GB tension increases. To counteract this increase in tension there must be a transport of atoms to the surface keeping the GB tension constant. This diffusion of atoms accounts for the constant surface tension in liquids. Then the argument,

$$\sigma_{GB} dA \text{ (work done)} = \gamma_{GB} dA \text{ (energy change)}$$

holds true. For solids, on the other hand, diffusion of atoms to the surface might not be sufficient and the surface tension can vary with an increase in surface area. For a solid, one can derive an expression for the change in Gibbs free energy, dG, upon the change of GB area, dA. dG is given by

$$\sigma_{GB} dA \text{ (work done)} = dG \text{ (energy change)} = \gamma_{GB} dA + A d\gamma_{GB}$$

which gives

$$\sigma_{GB} = \gamma_{GB} + \frac{A d\gamma_{GB}}{dA}$$

σ_{GB} is normally expressed in units of $\frac{N}{m}$ while γ_{GB} is normally expressed in units of $\frac{J}{m^2}$ $(J = Nm)$ since they are different physical properties.

Mechanical Equilibrium

In a two-dimensional isotropic material the grain boundary tension would be the same for the grains. This would give angle of 120° at GB junction where three grains meet. This would give the structure a hexagonal pattern which is the metastable state (or mechanical equilibrium) of the 2D specimen. A consequence of this is that to keep trying to be as close to the equilibrium as possible grains with fewer sides than six will bend the GB to try keep the 120° angle between each other. This results in a curved boundary with its curvature towards itself. A grain with six sides will as mentioned have straight boundaries while a grain with more than six sides will have curved boundaries with its curvature away from itself. A grain with six boundaries (i.e. hexagonal structure) are in a metastable state (i.e. local equilibrium) within the 2D structure. In three dimensions structural details are similar but much more complex and the metastable structure for a grain is a non-regular 14-sided polyhedra with doubly curved faces. In practice all arrays of grains are always unstable and thus always grows until its prevented by a counterforce.

Since the grains strive to minimize their energy and a curved boundary has a higher energy than a straight boundary. This means that the grain boundary will migrate towards the curvature. The consequence of this is that grains with less than 6 sides will decrease in size while grains with more than 6 sides will increase in size.

Grain growth happens due to motion of atoms across a grain boundary. Convex surfaces have a higher chemical potential than concave surfaces therefore grain boundaries will move toward their centre of curvature. As smaller particles tend to have a higher radius of curvature and this results in smaller grains losing atoms to larger grains and shrinking. This is a process called Ostwald ripening. Large grains grow at the expense of small grains. Grain growth in a simple model is found to follow:

$$G^m = G_0^m + Kt$$

Here G is final average grain size, G_0 is the initial average grain size, t is time, m is a factor between 2 and 4, and K is a factor given by:

$$K = K_0 e^{\frac{-Q}{RT}}$$

Here Q is the molar activation energy, R is the ideal gas constant, T is absolute temperature, and K_0 is a material dependent factor.

Reducing Grain Growth

Solute Ions: If a dopant is added to the material (example: Nd in $BaTiO_3$) the impurity will tend to stick to the grain boundaries. As the grain boundary tries to move (as atoms jump from the convex to concave surface) the change in concentration of the dopant at the grain boundary will impose a drag on the boundary. The original concentration of solute around the grain boundary will be asymmetrical in most cases. As the grain boundary tries to move the concentration on the side opposite of motion will have a higher concentration and therefore have a higher chemical potential. This increased chemical potential will act as a backforce to the original chemical potential gradient that is the reason for grain boundary movement. This decrease in net chemical potential will decrease the grain boundary velocity and therefore grain growth.

Fine Second Phase Particles

If particles of a second phase which are insoluble in the matrix phase are added to the powder in the form of a much finer powder than this will decrease grain boundary movement. When the grain boundary tries to move past the inclusion diffusion of atoms from one grain to the other will be hindered by the insoluble particle. Since it is beneficial for particles to reside in the grain boundaries and they exert a force in opposite direction compared to the grain boundary migration. This effect is called the Zener effect after the man who estimated this drag force to

$$F = \pi r \lambda \sin(2\theta)$$

where r is the radius of the particle and λ the interfacial energy of the boundary if there are N particles per unit volume their volume fraction f is

$$f = \frac{4}{3}\pi r^3 N$$

assuming they are randomly distributed. A boundary of unit area will intersect all particles within a volume of 2r which is 2Nr particles. So the number of particles n intersecting a unit area of grain boundary is:

$$n = \frac{3f}{2\pi r^2}$$

Now assuming that the grains only grow due to the influence of curvature, the driving force of growth is $\frac{2\lambda}{R}$ where (for homogeneous grain structure) R approximates to the mean diameter of the grains. With this the critical diameter that has to be reached before the grains ceases to grow:

$$nF_{max} = \frac{2\lambda}{D_{crit}}$$

This can be reduced to $D_{crit} = \frac{4r}{3f}$ so the critical diameter of the grains is dependent of the size and volume fraction of the particles at the grain boundaries.

It has also been shown that small bubbles or cavities can act as inclusion

More complicated interactions which slow grain boundary motion include interactions of the surface energies of the two grains and the inclusion and are discussed in detail by C.S. Smith.

Abnormal Grain Growth

Abnormal grain growth (AGG), also referred to as Exaggerated Grain Growth (EGG), is a grain growth phenomenon through which certain energetically favourable grains (crystallites) grow rapidly in a matrix of finer grains resulting in a bimodal grain size distribution. In ceramic materials this phenomena can result in the formation of elongated prismatic, acicular or needle-like grains in a densified matrix with implications for improved fracture toughness through the impedance of crack propagation.

Mechanisms

Abnormal grain growth is encountered in metallic or ceramic systems exhibiting one or more of several characteristics.

1. Secondary phase inclusions, precipitates or impurities above a certain threshold concentration.
2. High anisotropy in solid/liquid interfacial energy or grain boundary energy (solid/solid) in bulk materials.
3. Highly anisotropic surface energy in thin film materials.
4. High chemical inequilibrium.

Although many gaps remain in our fundamental understanding of AGG phenomena, in all cases abnormal grain growth occurrs as a result of very high local rates of interface migration and is enhanced by the localised formation of liquid at grain boundaries.

Significance

AGG is often recorded as an undesirable phenomenon occurring during the sintering of ceramic materials as rapidly growing grains may lower the hardness of the bulk material through Hall Petch type effects. However, the controlled introduction of dopants to bring about controlled AGG may be used to impart fibre-toughening in ceramic materials. In piezoelectric ceramics the occurrence of AGG may bring about the degradation of piezoelectric effect and thus in these systems AGG is avoided.

Example Systems

1. Rutile TiO_2 frequently exhibits a prismatic growth habit. In the presence of alkali dopants or a solid state $ZrSiO_4$ dopant, rutile has been observed to crystallise from a parent anatase phase material in the form of abnormally large grains existing in a matrix of finer equiaxed anatase or rutile grains.
2. Al_2O_3 with silica and/or yttria dopants/impurities has been reported to exhibit undesirable AGG.
3. $BaTiO_3$ barium titanate with an excess of TiO_2 is known to exhibit abnormal grain growth with profound consequences on this materials piezoelectric performance.
4. Tungsten carbide has been reported to exhibit AGG of faceted grains in the presence of a liquid cobalt-containing phase at grain boundaries
5. Silicon Nitride(Si_3N_4) may exhibit AGG depending on the size distribution of b-phase material in an a-Si_3N_4 precursor. This type of grain growth is of importance in the toughening of silicon nitride materials
6. Silicon carbide has been shown to exhibit improved fracture toughness as the result of AGG processes yielding elongated crack tip/wake bridging grains, with consequences for applications in ballistic armor. This type of crack-bridging based enhanced fracture toughness of ceramic materials exhibiting AGG is consistent with reported morphological effects on crack propagation in ceramics
6. Strontium Barium Niobate, used for electro-optics and dielectric applications is known to exhibit AGG with significant consequences on the material's electronic performance
7. $CaTiO_3$ perovskite systems doped with BaO have been observed to exhibit AGG without the formation of liquid as the result of polytype interfaces between solid phases

Direct Metal Laser Sintering

Direct metal laser sintering (DMLS) is an additive metal fabrication (Sintering) technology developed by EOS of Munich, Germany, often confused with the similar technical processes, selective laser sintering (SLS) or selective laser melting (SLM).

The process involves use of a 3D CAD model whereby a .stl file is created and sent to the machine's software. A technician works with this 3D model to properly orient the geometry for part building and adds supports structure as appropriate. Once this "build file" has been completed, it is "sliced" into the layer thickness the machine will build in and downloaded to the DMLS machine allowing the build to begin. The DMLS machine uses a high-powered 200 watt Yb-fibre optic laser. Inside the build chamber area, there is a material dispensing platform and a build platform along with a recoater blade used to move new powder over the build platform. The technology fuses metal powder into a solid part by melting it locally using the focused laser beam. Parts are built up additively layer by layer, typically using layers 20 micrometres thick. This process allows for highly complex geometries to be created directly from the 3D CAD data, fully automatically, in hours and without any tooling. DMLS is a net-shape process, producing parts with high accuracy and detail resolution, good surface quality and excellent mechanical properties.

Benefits

DMLS has many benefits over traditional manufacturing techniques. The ability to quickly produce a unique part is the most obvious because no special tooling is required and parts can be built in a matter of hours. Additionally, DMLS allows for more rigorous testing of prototypes. Since DMLS can use most alloys, prototypes can now be functional hardware made out of the same material as production components. DMLS is also one of the few additive manufacturing technologies being used in production. Since the components are built layer by layer, it is possible to design internal features and passages that could not be cast or otherwise machined. Complex geometries and assemblies with multiple components can be simplified to fewer parts with a more cost effective assembly. DMLS does not require special tooling like castings, so it is convenient for short production runs.

Applications

This technology is used to manufacture direct parts for a variety of industries including aerospace, dental, medical and other industries that have small to medium size, highly complex parts and the tooling industry to make direct tooling inserts. With a build envelop of 250 x 250 x 185 mm, and the ability to 'grow' multiple parts at one time, DMLS is a very cost and time effective technology. The technology is used both for rapid prototyping, as it decreases development time for new products, and production manufacturing as a cost saving method to simplify assemblies and complex geometries.

The Northwestern Polytechnical University of China is using a similar system to build structural titanium parts for aircraft. An EADS study shows that use of the process would reduce materials and waste in aerospace applications. On Sept. 5, 2013 Elon Musk tweeted an image of SpaceX's regeneratively cooled SuperDraco rocket engine chamber emerging from an EOS 3D metal printer, noting that it was composed of the Inconel superalloy.

Constraints

The aspects of size, feature details and surface finish, as well as print through error in the Z axis may be factors that should be considered prior to the use of the technology. However, by planning the build in the machine where most features are built in the x and y axis as the material is laid down, the feature tolerances can be managed well. Surfaces usually have to be polished to achieve mirror or extremely smooth finishes.

For production tooling, material density of a finished part or insert should be addressed prior to use. For example, in injection molding inserts, any surface imperfections will cause imperfections in the plastic part, and the inserts will have to mate with the base of the mold with temperature and surfaces to prevent problems. In this process metallic support structure removal and post processing of the part generated is a time consuming process and requires use of EDM and/or grinding machines having the same level of accuracy provided by the RP machine.

When using rapid prototyping machines, .stl files, which do not include anything but raw mesh data in binary (generated from Solid Works, CATIA, or other major CAD programs) need further conversion to .cli & .sli files (the format required for non stereolithography machines). Software converts .stl file to .sli files, as with the rest of the process, there can be costs associated with this step.

Materials

Currently available alloys used in the process include 17-4 and 15-5 stainless steel, maraging steel, cobalt chromium, inconel 625 and 718, and titanium Ti6Alv4. Theoretically, almost any alloy metal can be used in this process once fully developed and validated.

Capacitor Discharge Sintering

Capacitor discharge sintering (CDS) is an electric current assisted sintering (ECAS) technique. The technique is based on storage of electromagnetic energy in a high voltage capacitor bank, and discharge into the sintering apparatus at low voltage (<30 V) and high current through step-down transformers on a pre-compacted powder compact which is kept under pressure. The sintering mould and Electrodes are similar to those employed in field assisted sintering techniques (FAST) such as spark plasma sintering and single electromagnetic pulse sintering technologies.

The method, analogous to resistive sintering, is a direct evolution of a welding technology named Capacitor Discharge Welding (Kondensator-Impulsschweißen). CDS seems like an improvement of the less powerful capacitor discharge compaction patented by W.Knoess and M.Schlemmer (EP 0671232, US Patent 5529746).

Advantages of the technique are:

- short sintering time (less than one second)
- no need for a controlled atmosphere even during sintering of highly reactive materials such as titanium
- the ability to obtain metastable and nanocrystalline materials

Disambiguation

The technique has been studied by Element Six under the name of electro-discharge sintering. This name has been adopted by many authors in the past to describe a range of different technologies which typically adopt very high voltages and completely different machines. For this reason the technique which employs low voltages and high currents adapted from capacitor discharge welding has been named capacitor discharge sintering. Other authors also refer to this technology as spark plasma compaction (in reference to the well known Spark Plasma Sintering with whom it has in common only the use of electric currents).

Developments

Capacitor discharge sintering is at an experimental/research stage of development in Germany at the Ruhr-Universität Bochum where a prototype machine is installed. Industrialization and evolution of the technology is ongoing in EPoS srl, a start-up company from Torino. EPoS has patented a CDS based process which is claimed to produce components with improved properties at a competitive production cost. The company active since 2010 is employing solely proprietary technologies based on CDS to develop and produce abrasives, cemented carbide blanks, and a wide range of components produced with metallic alloys and composites.

Selective Laser Sintering, a Rapid Prototyping Technology

Selective laser sintering (SLS) is an additive manufacturing technique used for the low volume production of prototype models and functional components. Selective Laser Sintering uses lasers as its power source to sinter powdered material, binding it together to create a solid structure. It is often confused with another similar process called Selective Laser Melting (SLM), the difference being that it only sinters the powders together as opposed to achieving a full melt.

Selective Laser Sintering (SLS) was developed and patented by Dr. Carl Deckard and academic adviser, Dr. Joe Beaman at the University of Texas at Austin in the mid-1980s, under sponsorship of DARPA. Deckard and Beaman were involved in the resulting start up company DTM, established to design and build the Selective Laser Sintering Machines. In 2001, 3D Systems the biggest competitor of DTM and SLS technology acquired DTM. The most recent patent regarding Deckard's SLS technology was issued 28 January 1997 and expired 28 Jan 2014.

A similar process was patented without being commercialized by R. F. Housholder in 1979.

Technology

An additive manufacturing layer technology, SLS involves the use of a high power laser (for example, a carbon dioxide laser) to fuse small particles of plastic, metal (direct metal laser sintering), ceramic, or glass powders into a mass that has a desired three-dimensional shape. The laser selectively fuses powdered material by scanning cross-sections generated from a 3-D digital description of the part (for example from a CAD file or scan data) on the surface of a powder bed. After each cross-section is scanned, the powder bed is lowered by one layer thickness, a new layer of material is applied on top, and the process is repeated until the part is completed.

Because finished part density depends on peak laser power, rather than laser duration, a SLS machine typically uses a pulsed laser. The SLS machine preheats the bulk powder material in the powder bed somewhat below its melting point, to make it easier for the laser to raise the temperature of the selected regions the rest of the way to the melting point.

Unlike some other additive manufacturing processes, such as stereolithography (SLA) and fused deposition modelling (FDM), SLS does not require support structures due to the fact that the part being constructed is surrounded by unsintered powder at all times, this allows for the construction of previously impossible geometries.

Materials and Applications

Some SLS machines use single-component powder, such as direct metal laser sintering, powders are commonly produced by ball milling. However, most SLS machines use two-component powders, typically either coated powder or a powder mixture. In single-component powders, the laser melts only the outer surface of the particles (surface melting), fusing the solid non-melted cores to each other and to the previous layer.

Compared with other methods of additive manufacturing, SLS can produce parts from a relatively wide range of commercially available powder materials. These include polymers such as nylon (neat, glass-filled, or with other fillers) or polystyrene, metals including steel, titanium, alloy mixtures, and composites and green sand. The physical process can be full melting, partial melting, or liquid-phase sintering. Depending on the material, up to 100% density can be achieved with material properties comparable to those from conventional manufacturing methods. In many cases large numbers of parts can be packed within the powder bed, allowing very high productivity.

SLS technology is in wide use around the world due to its ability to easily make very complex geometries directly from digital CAD data. While it began as a way to build prototype parts early in the design cycle, it is increasingly being used in limited-run manufacturing to produce end-use parts. One less expected and rapidly growing application of SLS is its use in art.

Frit

A frit is a ceramic composition that has been fused in a special fusing oven, quenched to form a glass, and granulated. Frits form an important part of the batches used in compounding enamels and ceramic glazes; the purpose of this pre-fusion is to render any soluble and/or toxic components insoluble by causing them to combine with silica and other added oxides. However, not all glass that is fused and quenched in water is frit, as this method of cooling down very hot glass is also widely used in glass manufacture.

According to the OED, the origin of the word "frit" dates back to 1662 and is "a calcinated mixture of sand and fluxes ready to be melted in a crucible to make glass". Nowadays, the unheated raw materials of glass making are more commonly called "glass batch".

In antiquity, frit could be crushed to make pigments or shaped to create objects. It may also have served as an intermediate material in the manufacture of raw glass. The definition of frit tends to be variable and has proved a thorny issue for scholars. In recent centuries, frits have taken on a number of roles, such as biomaterials and additives to microwave dielectric ceramics. Frit in the form of alumino-silicate can be used in glaze free continuous casting refractories.

Frits are indispensable constituents of most industrial ceramic glazes which mature at temperatures below 1150°C. Frits are typically intermediates in the production of raw glass, as opposed to pigments and shaped objects. but they can be used as their own entities in a number of high-tech contexts. Frits made predominantly of silica, diboron trioxide (B_2O_3), and soda are used as enamels on steel pipes. Another type of frit can be used as a biomaterial, which is a material made to become a part of, or to come into intimate contact with, one or more living organisms. Molten soda-lime-silica glass can be "poured into water to obtain a frit", which is then ground to a powder. These powders can be used as "scaffolds for bone substitutions". Also, certain frits can be added to high-tech ceramics: such frits are made by milling zinc oxide (ZnO) and boric acid (H_3BO_3) with zirconium (Zr) beads, then heating this mixture to 1100°C, quenching it, and grinding it. This frit is then added to a lithium titanate (Li_2TiO_3) ceramic powder, which enables the ceramic to sinter at a lower temperature while still keeping its "microwave dielectric properties." Sintering is the high temperature fusing together of powder to form a solid, usually porous, material. In laboratory and industrial chemical process equipment, the term "frit" denotes a filter made by the sintering-together of glass particles to produce a piece of known porosity. The word is also used in a looser sense to denote a filter of *any* composition (for example, stainless steel, polyethylene) that is substituted where a frit of glass may otherwise have been used.

Yttria-Stabilized Zirconia

Yttria-stabilized zirconia (YSZ) is a ceramic in which the crystal structure of zirconium dioxide is made stable at room temperature by an addition of yttrium oxide. These oxides are commonly called "zirconia" (ZrO_2) and "yttria" (Y_2O_3), hence the name.

Stabilization

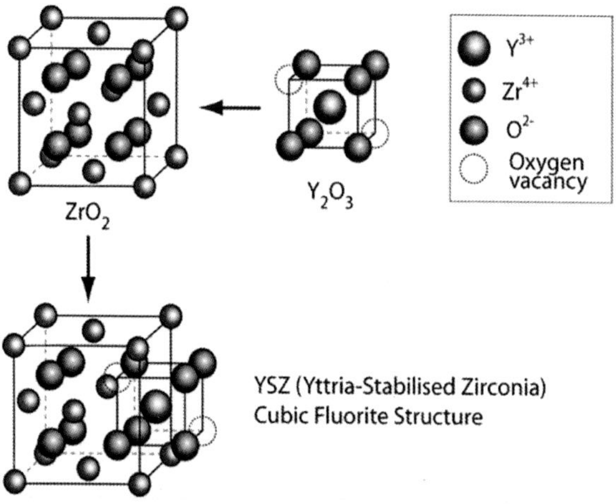

Figure: *Pure zirconium dioxide undergoes a phase transformations from monoclinic (stable at the room temperature) to tetragonal (at about 1000 °C) and then to cubic (at about 2370 °C), according to the scheme: monoclinic (1173 °C) ↔ tetragonal (2370 °C) ↔ cubic (2690 °C) ↔ melt*

Obtaining stable sintered zirconia ceramic products is difficult because of the large volume change accompanying the transition from tetragonal to monoclinic (about 9%). Stabilization of the cubic polymorph of zirconia over wider range of temperatures is accomplished by substitution of some of the Zr^{4+} ions (ionic radius of 0.82 Å, too small for ideal lattice of fluorite characteristic for the tetragonal zirconia) in the crystal lattice with slightly larger ions, e.g., those of Y^{3+} (ionic radius of 0.96 Å). The resulting doped zirconia materials are termed *stabilized zirconias.*

Materials related to YSZ include calcia-, magnesia-, ceria- or alumina-stabilized zirconias, or partially stabilized zirconias (PSZ). Hafnia stabilized Zirconia is also known.

Some of the abbreviations used in conjunction with stabilized zirconias are as follow:

- Partly stabilized zirconia ZrO_2:
  - o PSZ - *Partially Stabilized Zirconia*
  - o TZP -*Tetragonal Zirconia Polycrystal*
  - o 4YSZ: with 4 mol-% Y_2O_3 partially Stabilized ZrO_2, *Yttria Stabilized Zirconia*
- Fully stabilized zirconias ZrO_2:
  - o FSZ - *Fully Stabilized Zirconia*
  - o CSZ - *Cubic Stabilized Zirconia*
  - o 8YSZ - with 8 mol-% Y_2O_3 Fully Stabilized ZrO_2

Thermal Expansion Coefficient

The thermal expansion coefficients depends on the modification of zirconia as follows:

- Monoclinic: 7 · 10"6/K
- Tetragonal: 12 · 10"6/K
- Y_2O_3 stabilized: 10.5 · 10"6/K

Effect of Stabilization of Ionic Conductivity

The addition of yttria to pure zirconia replaces some of the Zr^{4+} ions in the zirconia lattice with Y^{3+} ions. This produces oxygen vacancies, as three O^{2-} ions replace four O^{2-} ions. It also permits yttrium stabilized zirconia to conduct O^{2-} ions (and thus conduct an electrical current), provided there is sufficient vacancy site mobility, a property that increases with temperature. This ability to conduct O^{2-} ions makes yttria-stabilized zirconia well suited to use in solid oxide fuel cells, although it requires that they operate at high enough temperatures.

The ionic conductivity of the stabilized zirconias increases with increasing dopant concentration (linearly for low dopant concentrations), then saturates, and then starts to decrease. The maximum ionic conductivity is obtained at Y_2O_3 concentration of about 8% (1000 °C).

Applications

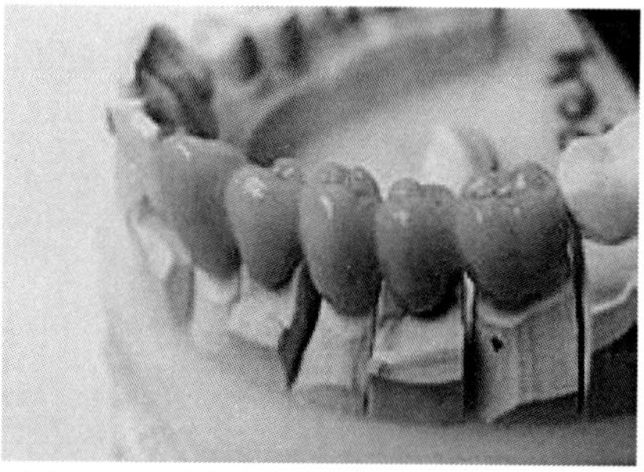

Figure: *Metal-free dental bridge*

YSZ has a number of applications:

- For its hardness and chemical inertness (e.g., tooth crowns).
- As a refractory (e.g., in jet engines).
- As a thermal barrier coating in gas turbines
- As an electroceramic due to its ion-conducting properties (e.g., to determine oxygen content in exhaust gases, to measure pH in high-temperature water, in fuel cells).
- Used in the production of a solid oxide fuel cell (SOFC). YSZ is used as the solid electrolyte, which enables oxygen ion conduction while blocking electronic conduction. In order to achieve sufficient ion conduction, an SOFC with a YSZ electrolyte must be operated at high temperatures (800 °C-1000 °C). While it is advantageous that YSZ retains mechanical robustness at those temperatures, overall the high temperature necessary is a disadvantage of current SOFCs, and is one of the major barriers to successful commercialization. The high density of YSZ is also necessary in order to physically separate the gaseous fuel from oxygen, or else the electrochemical system would produce no electrical power.
- For its hardness and optical properties in monocrystal form, it is used as jewellery.
- As a material for non-metallic knife blades, produced by Boker and Kyocera companies.
- In water-based pastes for *do-it-yourself* ceramics and cements. These contain microscopic YSZ milled fibres or sub-micrometre particles, often with potassium silicate and zirconium acetate binders (at mildly acidic pH). The cementation occurs on removal of water. The resulting ceramic material is suitable for very high temperature applications.
- YSZ doped with rare-earth materials can act as a thermographic phosphor and a luminescent material.

- Historically used for glowing rods in Nernst lamps.
- As a high precision alignment sleeve for optical fibre connector ferrules.

Metal Clay

Metal clay is a crafting medium consisting of very small particles of metal such as silver, gold, bronze, or copper mixed with an organic binder and water for use in making jewellery, beads and small sculptures. Originating in Japan in 1990, metal clay can be shaped just like any soft clay, by hand or using molds. After drying, the clay can be fired in a variety of ways such as in a kiln, with a handheld gas torch, or on a gas stove, depending on the type of clay and the metal in it. The binder burns away, leaving the pure sintered metal. Shrinkage of between 8% and 30% occurs (depending on the product used). Alloys such as bronze, sterling silver, and steel also are available.

Metal clay first came out in Japan in 1990 to allow craft jewellery makers to make sophisticated looking jewellery without the years of study needed to make fine jewellery.

Silver Metal Clay

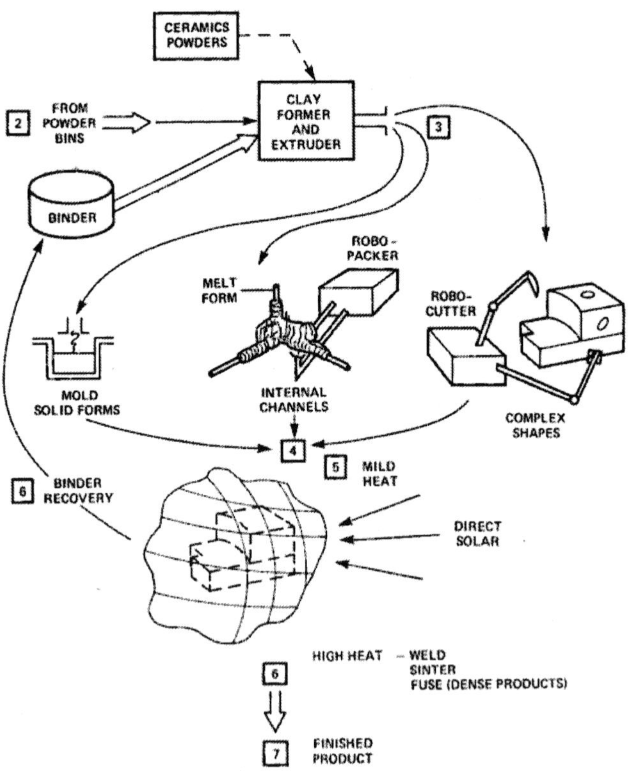

Figure: *Metal clay and pottery manufacturing*

Fine silver metal clay results in objects containing 99.9% pure silver, which is suitable for enameling. Gold metal clay is more expensive and provides richer colour. Lump metal clay is sold in sealed packets to keep it moist and workable. The silver versions are also available as a softer paste in a pre-filled syringe which can be used to produce extruded forms, in small jars of slip and as paper-like sheets, from which most of the moisture has been removed. Common brands of silver metal clay include Precious Metal Clay (PMC) and Art Clay Silver (ACS).

Precious Metal Clay (PMC)

PMC was developed in the early 1990s in Japan by metallurgist Masaki Morikawa. As a solid-phase sintered product of a precious metal powder used to form a precious metal article, the material consists of microscopic particles of pure silver or fine gold powder and a water-soluble, non-toxic, organic binder that burns off during firing. Success was first achieved with gold and later duplicated with silver. The PMC brand includes the following products:

- The original formula of PMC, now called "standard": fired at 900 °C (1,650 °F) for 2 hours, shrinks by 30% during firing.
- PMC+: fired at 900 °C (1,650 °F) for 10 minutes or 800 °C (1,470 °F) for 30 minutes; shrinks 15%, due to a particle size reduction. PMC+ is also available in sheet form which can be worked like paper; for example, for origami.
- PMC3: fired at 599 °C (1,110 °F) for 45 minutes or 699 °C (1,290 °F) for 10 minutes; shrinks by 10%. It can also be fired using a butane torch by heating it to orange heat for at least 2 minutes. It has a longer working life than the older formulations. It is also available in slip and paste forms which can be painted onto the surface of an object to be used as a mould.
- Aura 22: a 22 k gilding material, a gold paste intended to be painted onto the surface of silver PMC pieces, or ready-made silver objects.
- PMC Pro: a harder product which is only 0.900 silver, hence it cannot be hallmarked as sterling silver. It also requires kiln firing in a tub of activated carbon for 1 hour at 760 °C (1,400 °F).
- PMC Sterling: is fired at 815°C (1500°F) and shrinks by 10-15%. Because of the copper content in this formula, firing is a two-step process; step one is an open-shelf firing and step two requires a firing pan with activated carbon media.

Art Clay Silver (ACS)

ACS was developed by AIDA Chemical Industries, also a Japanese company. ACS followed PMC Standard with their Art Clay Original clay (more like

PMC+ than PMC Standard), which allows the user to fire with a handheld torch or on a gas hob. Owing to subtle differences in the binder and suggested firing times, this clay shrinks less than the PMC versions, approximately 8–10%.

Further developments introduced the Art Clay Slow Dry, a clay with a longer working time. Art Clay 650 and Art Clay 650 Slow Dry soon followed; both clays can be fired at 650 °C (1,202 °F), allowing the user to combine the clay with glass and sterling silver, which are affected negatively by the higher temperatures needed to fire the first generation clays. AIDA also manufacturers Oil Paste, a product used only on fired metal clay or milled fine silver, and Overlay Paste, which is designed for drawing designs on glass and porcelain.

In 2006 AIDA introduced the Art Clay Gold Paste, a more economical way to work with gold. The paste is painted onto the fired silver clay, then refired in a kiln, or with a torch or gas stove. When fired, it bonds with the silver, giving a 22ct gold accent. The same year also saw Art Clay Slow Tarnish introduced, a clay that tarnishes less rapidly than the other metal clays.

Lump Metal Clays

Lump metal clay in bronze was introduced in 2008 by Metal Adventures Inc. and in 2009 by Prometheus. Lump metal clays in copper were introduced in 2009 by Metal Adventures Inc. and Aida. Because of the lower cost, the bronze and copper metal clays are used by artists more often than the gold and silver metal clays in the American market place. Due to Hallmarking requirements laid out in the UK Bronze and Copper are not regarded as highly. The actual creation time of a bronze or copper piece is also far greater than that of its PMC3 counterpart.

Base Metal Clays

Base metal clays, such as bronze, copper, and steel metal clays are best fired in the absence of oxygen to eliminate the oxidation of copper by atmospheric oxygen. A simple means to accomplish this (place the pieces in activated carbon inside a container) was developed by Bill Struve.

4

Nanostructures of Metals and Ceramics

Generally, the physical properties of any given material can be characterized by some critical length, e.g., thermal diffusion length and attenuation distance. What makes nanoparticles very interesting and endows them with their remarkable properties is that their dimensions are smaller than a relevant critical length. Accordingly, the electron states of nanostructures are quantized, leading to new and usually striking electrical, thermal, magnetic, optical, and mechanical properties at the nanoscale. Accordingly, nanostructures are of both basic and practical interest since their physico-chemical properties can be tailored by controlling their size and shape at the nanoscale, leading to improved and/or novel applications.

The energy spectrum (i.e., the ensemble of discrete eigenenergies) of a quantum well, quantum wire, or quantum dot can be engineered by controlling (i) the size and shape of the confinement region and (ii) the strength of the confinement potential. The resulting control over the physico-chemical properties of the nanostructures is limited only by the accuracy of the experimental techniques used for the fabrication of the low-dimensional structures. The situation is not unlike that of quantum phenomena, many of which were described at the beginning of the 20th century but could not be demonstrated until the 1960s to 1970s when appropriate nanofabrication techniques were developed.

Nanostructure of Metals

The main reason for interest in nanometals between 1 to 50 nm in size is their high surface areas. It is estimated that 10% of metal atoms are on the surface for a particle with a 10 nm diameter. In comparison, 60% of the atoms are on the surface for a 2.5 nm size particle. Having such a high percentage of atoms exposed at the surface gives nanometals a distinct advantage over bulk materials.

Irilliant Advantage – High Volume, Low Cost Production

Currently, most nanometals are produced in small quantities by reduction of metal salts in water or organic solutions. Obtained particles have decent size distribution, but the size is typically large (>20 nm) and not well controlled. Some nanometals are produced on a larger, industrial scale in the gas phase by vacuum evaporation techniques and then stabilized in solution using standard surfactants. In this case, particles tend to agglomerate and the shapes and the sizes of nanoparticles are not well controlled.

Nanostructure of Ceramic

A ceramic is an inorganic, nonmetallic solid prepared by the action of heat and subsequent cooling. Ceramic materials may have a crystalline or partly crystalline structure, or may be amorphous (e.g., a glass). Because most common ceramics are crystalline, the definition of ceramic is often restricted to inorganic crystalline materials, as opposed to the noncrystalline glasses, a distinction followed here.

The earliest ceramics made by humans were pottery objects, including 27,000 year old figurines, made from clay, either by itself or mixed with other materials, hardened in fire. Later ceramics were glazed and fired to create a coloured, smooth surface. Ceramics now include domestic, industrial and building products and a wide range of ceramic art. In the 20th century, new ceramic materials were developed for use in advanced ceramic engineering; for example, in semiconductors.

The word "ceramic" comes from the Greek word êåñáìéêüò (*keramikos*), "of pottery" or "for pottery", from êÝñáìïò (*keramos*), "potter's clay, tile, pottery". The earliest mention of the root "ceram-" is the Mycenaean Greek *ke-ra-me-we*, "workers of ceramics", written in Linear B syllabic script. "Ceramic" may be used as an adjective describing a material, product or process; or as a singular noun, or, more commonly, as a plural noun, "ceramics".

Types of Ceramic Product

For convenience, ceramic products are usually divided into four sectors; these are shown below with some examples:

- Structural, including bricks, pipes, floor and roof tiles
- Refractories, such as kiln linings, gas fire radiants, steel and glass making crucibles

- White Wares, including tableware, cookware, wall tiles, pottery products and sanitary ware
- Technical, is also known as engineering, advanced, special, and in Japan, fine ceramics. Such items include tiles used in the Space Shuttle program, gas burner nozzles, ballistic protection, nuclear fuel uranium oxide pellets, biomedical implants, coatings of jet engine turbine blades, ceramic disk brake, missile nose cones, bearing (mechanical), etc. Frequently, the raw materials do not include clays.

Examples of White Ware Ceramics

- Earthenware, which is often made from clay, quartz and feldspar.
- Stoneware
- Porcelain, which is often made from kaolin
- Bone china

Classification of Technical Ceramics

Technical ceramics can also be classified into three distinct material categories:

- Oxides: alumina, beryllia, ceria, zirconia
- Nonoxides: carbide, boride, nitride, silicide
- Composite materials: particulate reinforced, fibre reinforced, combinations of oxides and nonoxides.

Each one of these classes can develop unique material properties because ceramics tend to be crystalline.

Other Applications of Ceramics

- Knife blades: the blade of a ceramic knife will stay sharp for much longer than that of a steel knife, although it is more brittle and can snap from a fall onto a hard surface.
- Ceramic brake disks for vehicles are resistant to abrasion at high temperatures.
- Advanced composite ceramic and metal matrices have been designed for most modern armoured fighting vehicles because they offer superior penetrating resistance against shaped charges (such as HEAT rounds) and kinetic energy penetrators.
- Ceramics such as alumina and boron carbide have been used in ballistic armored vests to repel large-caliber rifle fire. Such plates are known commonly as small arms protective inserts, or SAPIs. Similar material is used to protect the cockpits of some military airplanes, because of the low weight of the material.

- Ceramics can be used in place of steel for ball bearings. Their higher hardness means they are much less susceptible to wear and typically last for triple the lifetime of a steel part. They also deform less under load, meaning they have less contact with the bearing retainer walls and can roll faster. In very high speed applications, heat from friction during rolling can cause problems for metal bearings, which are reduced by the use of ceramics. Ceramics are also more chemically resistant and can be used in wet environments where steel bearings would rust. In some cases, their electricity-insulating properties may also be valuable in bearings. Two drawbacks to ceramic bearings are a significantly higher cost and susceptibility to damage under shock loads.
- In the early 1980s, Toyota researched production of an adiabatic engine using ceramic components in the hot gas area. The ceramics would have allowed temperatures of over 3000°F (1650°C). The expected advantages would have been lighter materials and a smaller cooling system (or no need for one at all), leading to a major weight reduction. The expected increase of fuel efficiency of the engine (caused by the higher temperature, as shown by Carnot's theorem) could not be verified experimentally; it was found that the heat transfer on the hot ceramic cylinder walls was higher than the transfer to a cooler metal wall as the cooler gas film on the metal surface works as a thermal insulator. Thus, despite all of these desirable properties, such engines have not succeeded in production because of costs for the ceramic components and the limited advantages. (Small imperfections in the ceramic material with its low fracture toughness lead to cracks, which can lead to potentially dangerous equipment failure.) Such engines are possible in laboratory settings, but mass production is not feasible with current technology.
- Work is being done in developing ceramic parts for gas turbine engines. Currently, even blades made of advanced metal alloys used in the engines' hot section require cooling and careful limiting of operating temperatures. Turbine engines made with ceramics could operate more efficiently, giving aircraft greater range and payload for a set amount of fuel.
- Recent advances have been made in ceramics which include bioceramics, such as dental implants and synthetic bones. Hydroxyapatite, the natural mineral component of bone, has been made synthetically from a number of biological and chemical sources and can be formed into ceramic materials. Orthopedic implants coated with these materials bond readily to bone and other tissues in the body without rejection or inflammatory reactions so are of great interest for gene delivery and tissue engineering scaffolds. Most hydroxyapatite ceramics are very porous and lack

mechanical strength, and are used to coat metal orthopedic devices to aid in forming a bond to bone or as bone fillers. They are also used as fillers for orthopedic plastic screws to aid in reducing the inflammation and increase absorption of these plastic materials. Work is being done to make strong, fully dense nanocrystalline hydroxyapatite ceramic materials for orthopedic weight bearing devices, replacing foreign metal and plastic orthopedic materials with a synthetic, but naturally occurring, bone mineral. Ultimately, these ceramic materials may be used as bone replacements or with the incorporation of protein collagens, synthetic bones.

- High-tech ceramic is used in watchmaking for producing watch cases. The material is valued by watchmakers for its light weight, scratch resistance, durability and smooth touch. IWC is one of the brands that initiated the use of ceramic in watchmaking. The case of the IWC 2007 Top Gun edition of the Pilot's Watch double chronograph is crafted in black ceramic.

Types of Ceramic Material

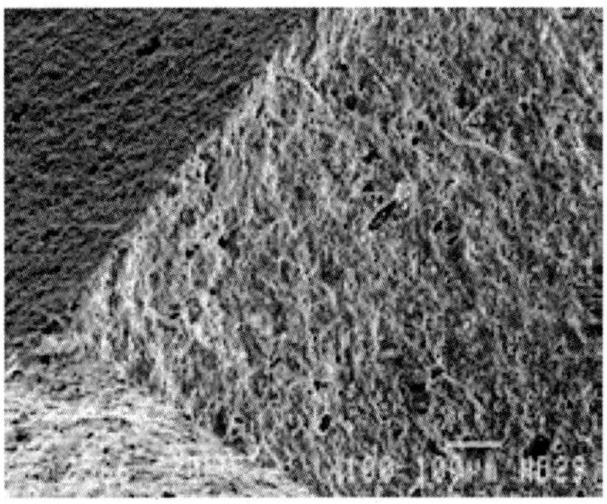

Figure: *A low magnification SEM micrograph of an advanced ceramic material. The properties of ceramics make fracturing an important inspection method.*

A ceramic material is an inorganic, non-metallic, often crystalline oxide, nitride or carbide material. Some elements, such as carbon or silicon, may be considered ceramics. Ceramic materials are brittle, hard, strong in compression, weak in shearing and tension. They withstand chemical erosion that occurs in other materials subjected to acidic or caustic environments. Ceramics generally can withstand very high temperatures, such as temperatures that range from 1,000 °C to 1,600 °C (1,800 °F to 3,000 °F). A glass is often not understood as a ceramic because of its amorphous (noncrystalline) character. However, glassmaking involves several steps of the ceramic process and its mechanical properties are similar to ceramic materials.

Traditional ceramic raw materials include clay minerals such as kaolinite, whereas more recent materials include aluminium oxide, more commonly known as alumina. The modern ceramic materials, which are classified as advanced ceramics, include silicon carbide and tungsten carbide. Both are valued for their abrasion resistance, and hence find use in applications such as the wear plates of crushing equipment in mining operations. Advanced ceramics are also used in the medicine, electrical and electronics industries.

Crystalline Ceramics

Crystalline ceramic materials are not amenable to a great range of processing. Methods for dealing with them tend to fall into one of two categories – either make the ceramic in the desired shape, by reaction *in situ*, or by "forming" powders into the desired shape, and then sintering to form a solid body. Ceramic forming techniques include shaping by hand (sometimes including a rotation process called "throwing"), slip casting, tape casting (used for making very thin ceramic capacitors, e.g.), injection molding, dry pressing, and other variations. Details of these processes are described in the two books listed below. A few methods use a hybrid between the two approaches.

Noncrystalline Ceramics

Noncrystalline ceramics, being glass, tend to be formed from melts. The glass is shaped when either fully molten, by casting, or when in a state of toffee-like viscosity, by methods such as blowing into a mold. If later heat treatments cause this glass to become partly crystalline, the resulting material is known as a glass-ceramic, widely used as cook-top and also as a glass composite material for nuclear waste disposal.

Ceramics in Archaeology

Ceramic artifacts have an important role in archaeology for understanding the culture, technology and behaviour of peoples of the past. They are among the most common artifacts to be found at an archaeological site, generally in the form of small fragments of broken pottery called sherds. Processing of collected sherds can be consistent with two main types of analysis: technical and traditional. Traditional analysis involves sorting ceramic artifacts, sherds and larger fragments into specific types based on style, composition, manufacturing and morphology. By creating these typologies it is possible to distinguish between different cultural styles, the purpose of the ceramic and technological state of the people among other conclusions. In addition, by looking at stylistic changes of ceramics over time is it possible to separate (seriate) the ceramics into distinct diagnostic groups (assemblages). A comparison of ceramic artifacts with known dated assemblages allows for a chronological assignment of these pieces.

The technical approach to ceramic analysis involves a finer examination of the composition of ceramic artifacts and sherds to determine the source of the material and through this the possible manufacturing site. Key criteria are the composition of the clay and the temper used in the manufacture of the article under study: temper is a material added to the clay during the initial production stage, and it is used to aid the subsequent drying process. Types of temper include shell pieces, granite fragments and ground sherd pieces called 'grog'. Temper is usually identified by microscopic examination of the temper material. Clay identification is determined by a process of refiring the ceramic, and assigning a colour to it using Munsell Soil Colour notation. By estimating both the clay and temper compositions, and locating a region where both are known to occur, an assignment of the material source can be made. From the source assignment of the artifact further investigations can be made into the site of manufacture.

Characterization by Scattering Techniques and EXAFS

X-ray Absorption Spectroscopy (XAS) includes both Extended X-Ray Absorption Fine Structure (EXAFS) and X-ray Absorption Near Edge Structure (XANES). XAS is the measurement of the x-ray absorption coefficient ($\mu(E)$ in the equations below) of a material as a function of energy. X-rays of a narrow energy resolution are shone on the sample and the incident and transmitted x-ray intensity is recorded as the incident x-ray energy is incremented. The number of x-ray photons that are transmitted through a sample (I_t) is equal to the number of x-ray photons shone on the sample (I_0) multiplied by a decreasing exponential that depends on the type of atoms in the sample, the absorption coefficient μ, and the thickness of the sample x.

$$I_t = I_0 e^{-\mu x}$$

The absorption coefficient is obtained by taking the log ratio of the incident x-ray intensity to the transmitted x-ray intensity.

$$\mu = \frac{-ln(I_t / I_0)}{x}$$

When the incident x-ray energy matches the binding energy of an electron of an atom within the sample, the number of x-rays absorbed by the sample increases dramatically, causing a drop in the transmitted x-ray intensity. This results in an absorption edge. Each element on the periodic table has a set of unique absorption edges corresponding to different binding energies of its electrons. This gives XAS element selectivity. XAS spectra are most often collected at synchrotrons. Because X-rays are highly penetrating, XAS samples can be gases, solids or liquids. And because of the brilliance of synchrotron X-ray sources the concentration of the absorbing element can be as low as a few ppm.

EXAFS spectra are displayed as graphs of the absorption coefficient of a given material versus energy, typically in a 500 – 1000 eV range beginning before an absorption edge of an element in the sample. The x-ray absorption coefficient is usually normalized to unit step height. This is done by regressing a line to the region before and after the absorption edge, subtracting the pre-edge line from the entire data set and dividing by the absorption step height, which is determined by the difference between the pre-edge and post-edge lines at the value of E0 (on the absorption edge).

The normalized absorption spectra are often called XANES spectra. These spectra can be used to determine the average oxidation state of the element in the sample. The XANES spectra are also sensitive to the coordination environment of the absorbing atom in the sample. Finger printing methods have been used to match the XANES spectra of an unknown sample to those of known "standards". Linear combination fitting of several different standard spectra can give an estimate to the amount of each of the known standard spectra within an unknown sample.

X-ray absorption spectra are produced over the range of 200 – 35,000 eV. The dominant physical process is one where the absorbed photon ejects a core photoelectron from the absorbing atom, leaving behind a core hole. The atom with the core hole is now excited. The ejected photoelectron's energy will be equal to that of the absorbed photon minus the binding energy of the initial core state. The ejected photoelectron interacts with electrons in the surrounding non-excited atoms. If the ejected photoelectron is taken to have a wave-like nature and the surrounding atoms are described as point scatterers, it is possible to imagine the backscattered electron waves interfering with the forward-propagating waves. The resulting interference pattern shows up as a modulation of the measured absorption coefficient, thereby causing the oscillation in the EXAFS spectra. A simplified plane-wave single-scattering theory has been used for interpretation of EXAFS spectra for many years, although modern methods (like FEFF, GNXAS) have shown that curved-wave corrections and multiple-scattering effects can not be neglected. The photelectron scattering amplitude in the low energy range (5-200 eV) of the photoelectron kinetic energy become much larger so that multiple scattering events become dominant in the NEXAFS (or XANES) spectra.

The wavelength of the photoelectron is dependent on the energy and phase of the backscattered wave which exists at the central atom. The wavelength changes as a function of the energy of the incoming photon. The phase and amplitude of the backscattered wave are dependent on the type of atom doing the backscattering and the distance of the backscattering atom from the central atom. The dependence of the scattering on atomic species makes it possible to obtain information pertaining to the chemical coordination environment of the original absorbing (centrally excited) atom by analyzing these EXAFS data.

Experimental Considerations

Since EXAFS requires a tunable x-ray source, data are always collected at synchrotrons, often at beamlines which are especially optimized for the purpose. The utility of a particular synchrotron to study a particular solid depends on the brightness of the x-ray flux at the absorption edges of the relevant elements.

Applications

XAS is an interdisciplinary technique and its unique properties, as compared to x-ray diffraction, have been exploited for understanding the details of local structure in:

- glass, amorphous and liquid systems
- solid solutions
- Doping and ionic implantation materials for electronics
- local distortions of crystal lattices
- organometallic compounds
- metalloproteins
- metal clusters
- vibrationaldynamics
- ions in solutions
- speciation of elements

Example of Significance

EXAFS is, like NEXAFS/XANES, a highly sensitive technique with elemental specificity. As such, EXAFS is an extremely useful way to determine the chemical state of practically important species which occur in very low abundance or concentration. Frequent use of EXAFS occurs in environmental chemistry, where scientists try to understand the propagation of pollutants through an ecosystem. EXAFS can be used along with accelerator mass spectrometry in forensic examinations, particularly in nuclear non-proliferation applications.

For an example of an EXAFS study of uranium chemistry in glass see, and for a general study of trivalent lanthanides and actinides in chloride containing aqueous media can be read at.

Synchrotron

A synchrotron is a particular type of cyclic particle accelerator originating from the cyclotron in which the guiding magnetic field (bending the particles into a closed path) is time-dependent, being *synchronized* to a particle beam of increasing kinetic energy. The synchrotron is one of the first accelerator concepts to enable the construction of large-scale facilities, since bending, beam focusing and acceleration can be separated into different components. Edwin McMillan constructed the first electron synchrotron in 1945, although Vladimir Veksler had already (unknown to McMillan) published the principle in a Soviet journal in 1944. The first proton synchrotron was designed by Sir Marcus Oliphant and built in 1952.

Differentiation

A storage ring is a special type of synchrotron in which the kinetic energy of the particles is kept constant. A synchrotron light source is a combination of different electron accelerator types, including a storage ring in which the desired electromagnetic radiation is generated. This radiation is then used in experimental stations located on different beamlines. In addition to the storage ring, a synchrotron light source usually contains a linear accelerator (linac) and another synchrotron which is sometimes called *booster* in this context. The linac and the booster are used to successively accelerate the electrons to their final energy before they are magnetically "kicked" into the storage ring. Synchrotron light sources in their entirety are sometimes called "synchrotrons", although this is technically incorrect. A cyclic collider is also a combination of different accelerator types, including two intersecting storage rings and the respective pre-accelerators.

Principle of Operation

While a classical cyclotron uses both a constant guiding magnetic field and a constant-frequency electromagnetic field (and is working in classical approximation), its successor, the isochronous cyclotron, works by local variations of the guiding magnetic field, adapting the increasing relativistic mass of particles during acceleration.

Figure: *A drawing of the Cosmotron*

In a synchrotron, this adaptation is done by variation of the magnetic field strength in time, rather than in space. For particles that are not close to the speed of light, the frequency of the applied electromagnetic field may also change to follow their non-constant circulation time. By increasing these parameters accordingly as the particles gain energy, their circulation path can be held constant as they are accelerated. This allows the vacuum chamber for the particles to be a large thin torus, rather than a disk as in previous, compact accelerator designs. Also, the thin profile of the vacuum chamber allowed for a more efficient use of magnetic fields than in a cyclotron, enabling the cost-effective construction of larger synchrotrons.

While the first synchrotrons and storage rings like the Cosmotron and ADA strictly used the toroid shape, the strong focusing principle independently discovered by Ernest Courant et al. and Nicholas Christofilos allowed the complete separation of the accelerator into components with specialised functions along the particle path, shaping the path into a round-cornered polygon. Some important components are given by radio frequency cavities for direct acceleration, dipole magnets (*bending magnets*) for deflection of particles (to close the path), and quadrupole / sextupole magnets for beam focusing.

The combination of time-dependent guiding magnetic fields and the strong focusing principle enabled the design and operation of modern large-scale accelerator facilities like colliders and synchrotron light sources. The straight sections along the closed path in such facilities are not only required for radio frequency cavities, but also for particle detectors (in colliders) and photon generation devices such as wigglers and undulators (in third generation synchrotron light sources).

The maximum energy that a cyclic accelerator can impart is typically limited by the maximum strength of the magnetic fields and the minimum radius (maximum curvature) of the particle path. Thus one method for increasing the energy limit is to use superconducting magnets, these not being limited by magnetic saturation. Electron/positron accelerators may also be limited by the emission of synchrotron radiation, resulting in a partial loss of the particle beam's kinetic energy. The limiting beam energy is reached when the energy lost to the lateral acceleration required to maintain the beam path in a circle equals the energy added each cycle. More powerful accelerators are built by using large radius paths and by using more numerous and more powerful microwave cavities. Lighter particles (such as electrons) lose a larger fraction of their energy when deflected. Practically speaking, the energy of electron/positron accelerators is limited by this radiation loss, while this does not play a significant role in the dynamics of proton or ion accelerators. The energy of such accelerators is limited strictly by the strength of magnets and by the cost.

Injection Procedure

Unlike in a cyclotron, synchrotrons are unable to accelerate particles from zero kinetic energy; one of the obvious reasons for this is that its closed particle path would be cut by a device that emits particles. Thus, schemes were developed to inject pre-accelerated particle beams into a synchrotron. The pre-acceleration can be realised by a chain of other accelerator structures like a linac, a microtron or another synchrotron; all of these in turn need to be fed by a particle source comprising a simple high voltage power supply, typically a Cockcroft-Walton generator.

Starting from an appropriate initial value determined by the injection energy, the field strength of the dipole magnets is then increased. If the high energy particle are emitted at the end of the acceleration procedure, e.g. to a target or to another accelerator, the field strength is again decreased to injection level, starting a new *injection cycle*. Depending on the method of magnet control used, the time interval for one cycle can vary substantially between different installations.

In Large-Scale Facilities

One of the early large synchrotrons, now retired, is the Bevatron, constructed in 1950 at the Lawrence Berkeley Laboratory. The name of this proton accelerator comes from its power, in the range of 6.3 GeV (then called BeV for billion electron volts; the name predates the adoption of the SI prefix giga-). A number of transuranium elements, unseen in the natural world, were first created with this machine. This site is also the location of one of the first large bubble chambers used to examine the results of the atomic collisions produced here.

Another early large synchrotron is the Cosmotron built at Brookhaven National Laboratory which reached 3.3 GeV in 1953.

As Part of Colliders

Until August 2008, the highest energy collider in the world was the Tevatron, at the Fermi National Accelerator Laboratory, in the United States. It accelerated protons and antiprotons to slightly less than 1 TeV of kinetic energy and collided them together. The Large Hadron Collider (LHC), which has been built at the European Laboratory for High Energy Physics (CERN), has roughly seven times this energy (so proton-proton collisions occur at roughly 14 TeV). It is housed in the 27 km tunnel which formerly housed the Large Electron Positron (LEP) collider, so it will maintain the claim as the largest scientific device ever built. The LHC will also accelerate heavy ions (such as lead) up to an energy of 1.15 PeV.

The largest device of this type seriously proposed was the Superconducting Super Collider (SSC), which was to be built in the United States. This design, like others, used superconducting magnets which allow more intense magnetic fields to be created without the limitations of core saturation. While construction was begun, the project was cancelled in 1994, citing excessive budget overruns — this was due to naïve cost estimation and economic management issues rather than any basic engineering flaws. It can also be argued that the end of the Cold War resulted in a change of scientific funding priorities that contributed to its ultimate cancellation. However, the tunnel built for its placement still remains, although empty. While there is still potential for yet more powerful proton and heavy particle cyclic accelerators, it appears that the next step up in electron beam energy must avoid losses due to synchrotron radiation. This will require a return to the linear accelerator, but with devices significantly longer than those currently in use. There is at present a major effort to design and build the International Linear Collider (ILC), which will consist of two opposing linear accelerators, one for electrons and one for positrons. These will collide at a total centre of mass energy of 0.5 TeV.

As Part of Synchrotron Light Sources

Synchrotron radiation also has a wide range of applications and many 2nd and 3rd generation synchrotrons have been built especially to harness it. The largest of those 3rd generation synchrotron light sources are the European Synchrotron Radiation Facility (ESRF) in Grenoble, France, the Advanced Photon Source (APS) near Chicago, USA, and SPring-8 in Japan, accelerating electrons up to 6, 7 and 8 GeV, respectively.

Synchrotrons which are useful for cutting edge research are large machines, costing tens or hundreds of millions of dollars to construct, and each beamline (there may be 20 to 50 at a large synchrotron) costs another two or three million dollars on average. These installations are mostly built by the science funding agencies of governments of developed countries, or by collaborations between several countries in a region, and operated as infrastructure facilities available to scientists from universities and research organisations throughout the country, region, or world. More compact models, however, have been developed, such as the Compact Light Source.

X-ray Absorption Spectroscopy

X-ray absorption spectroscopy (XAS) is a widely used technique for determining the local geometric and/or electronic structure of matter. The experiment is usually performed at synchrotron radiation sources, which provide intense and tunable X-ray beams. Samples can be in the gas-phase, solution, or condensed matter (i.e. solids).

XAS data are obtained by tuning the photon energy using a crystalline monochromator to a range where core electrons can be excited (0.1-100 keV photon energy). The "name" of the edge depends upon the core electron which is excited: the principal quantum numbers n=1, 2, and 3, correspond to the K-, L-, and M-edges, respectively. For instance, excitation of a 1s electron occurs at the K-edge, while excitation of a 2s or 2p electron occurs at an L-edge.

There are three main regions found on a spectrum generated by XAS data. 1) The "absorption threshold" determined by the transition to the lowest unoccupied states : a) the states at the Fermi energy in metals giving a "rising edge" with an invtan shape; b) the bound core excitons in insulators with a Lorentzian line-shape (they occur in a pre-edge region at energies lower than the transitions to the lowest unoccupied level); 2) The X-ray Absorption Near-Edge Structure XANES introduced in 1980 and later in 1983 called also NEXAFS (Near-edge X-ray Absorption Fine Structure) which are dominated by core transitions to quasi bound states (multiple scattering resonances) for photoelectrons with kinetic energy in the range from 10 to 150 eV above the chemical potential, called "shape resonances" in molecular spectra since they are due to final states of short life-time degenerate with the continuum with the Fano line-shape. In this range multi-electron exitations and many-body final states in strongly correlated systems are relevant; 3) In the high kinetic energy range of the photoelectron the scattering cross-section with neighbour atoms is weak and the absorption spectra are dominated by EXAFS (Extended X-ray Absorption Fine Structure) where the scattering of the ejected photoelectron off neighbouring atoms can be approximated by single scattering events. After it was shown in 1985 that multiple scattering theory can interpret both XANES and EXAFS the experimental analyis focusin on both regions is called XAFS. XAS is a type of absorption spectroscopy from a core initial state with a well defined symmetry therefore the quantum mechanical selection rules select the symmetry of the final states in the continuum which usually are mixture of multiple components. The most intense features are due to electric-dipole allowed transitions (i.e. $\Delta \ell = \pm 1$) to unoccupied final states. For example, the most intense features of a K-edge are due to core transitions from 1s $\rightarrow$ p-like final states, while the most intense features of the L_3-edge are due to 2p $\rightarrow$ d-like final states.

XAS methodology can be broadly divided into four experimental categories that can give complementary results to each other: Metal K-edge, metal L-edge, ligand K-edge, and EXAFS.

Xanes/Nexafs

X-ray Absorption Near Edge Structure (XANES), also known as Near edge X-ray absorption fine structure (NEXAFS) is a type of absorption spectroscopy .

XANES indicates the features in the X-ray Absorption Spectra (XAS) of condensed matter due to the photoabsorption cross section for electronic transitions from an atomic core level to final states in the energy region of 50-100 eV above the selected atomic core level ionization potential where the wavelength of the photoelectron is larger than the interatomic distance between the absorbing atom and its first neighbour atoms.

Both XANES and NEXAFS are acceptable terms for the same technique. XANES name was invented in 1980 by Antonio Bianconi to indicate strong absorption peaks in x-ray absorption spectra in condensed matter due to multiple scattering resonances above the ionization potential. The XANES energy range above the ionization potential NEXAFS, introduced in 1983 by Jo Stohr is synonymous with XANES, is generally used when applied to surface and molecular science

Theory

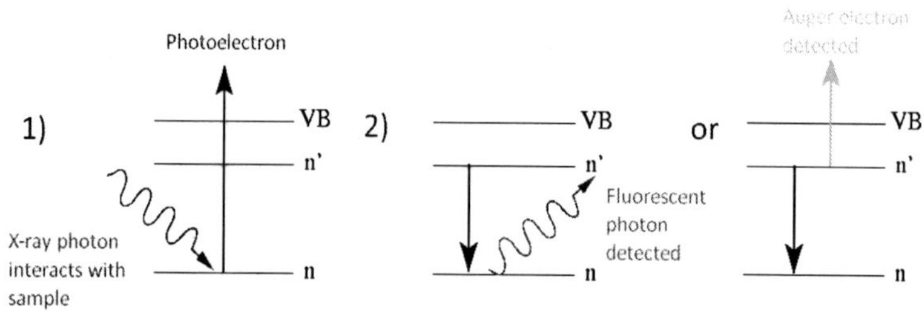

Figure: *The fundamental processes which contribute to XANES spectra: 1) photoabsorption of an x-ray into a core level followed by photoelectron emission, followed by either 2) (left) filling of the core hole by an electron in another level, accompanied by fluorescence; or (right) filling of the core hole by an electron in another level followed by emission of an Auger electron.*

The fundamental phenomenon underlying XANES is the absorption of an x-ray photon by condensed matter with the formation of many body excited states characterized by a core hole in a selected atomic core level. In the single particle theory approximation the system is separated into one electron in the core levels of the selected atomic species of the system and N-1 passive electrons. In this approximation the final state is described by a core hole in the atomic core level and an excited photoelectron. The final state has a very short life time because of the short life-time of the core hole and the short mean free path of the excited photoelectron with kinetic energy in the range around 20-50 eV. The core hole is filled either via an Auger process or by capture of an electron from another shell followed by emission of a fluorescent photon. The difference between NEXAFS and traditional photoemission experiments is that in photoemission, the initial photoelectron itself is measured, while in NEXAFS the fluorescent photon or Auger electron or an inelastically scattered photoelectron may also be measured. The distinction sounds trivial but is actually significant: in photoemission the final state of the emitted electron captured in the detector must be an extended, free-electron state. By contrast in NEXAFS the final state of the photoelectron may be a bound state such as an exciton since the photoelectron itself need not be detected. The effect of measuring fluorescent photons, Auger electrons, and directly emitted electrons is to sum over all possible final states of the photoelectrons, meaning that what NEXAFS measures is the total joint density of states of the initial core level with all final states, consistent with conservation rules. The distinction is critical because in spectroscopy final states are more susceptible to many-body effects than initial states, meaning that NEXAFS spectra are more easily calculable than photoemission spectra. Due to the summation over final states, various sum rules are helpful in the interpretation of NEXAFS spectra. When the x-ray photon energy resonantly connects a core level with a narrow final state in a solid, such as an exciton, readily identifiable characteristic peaks will appear in the spectrum. These narrow characteristic spectral peaks give the NEXAFS technique a lot of its analytical power as illustrated by the B 1s π* exciton shown in the second Figure.

Synchrotron radiation has a natural polarization that can be utilised to great advantage in NEXAFS studies. The commonly studied molecular adsorbates have sigma and pi bonds that may have a particular orientation on a surface. The angle dependence of the x-ray absorption tracks the orientation of resonant bonds due to dipole selection rules.

Experimental Considerations

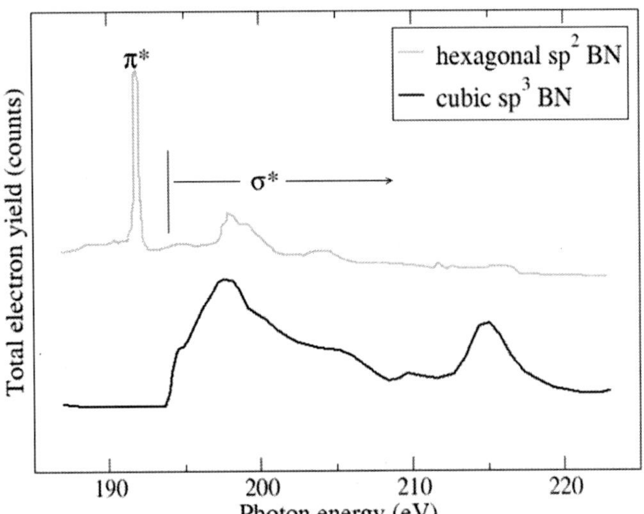

Figure: *Normal-incidence boron 1s x-ray absorption spectra for two types of BN powder. The cubic phase shows only s-bonding while the hexagonal phase shows both p and s bonding.*

Soft x-ray aborption spectra are usually measured either through the *fluorescent yield,* in which emitted photons are monitored, or *total electron yield,* in which the sample is connected to ground through an ammeter and the neutralization current is monitored. Because XANES measurements require an intense tunable source of soft x-rays, they are performed at synchrotrons. Because soft x-rays are absorbed by air, the synchrotron radiation travels from the ring in an evacuated beam-line to the end-station where the specimen to be studied is mounted. Specialised beam-lines intended for XANES studies often have additional capabilities such as heating a sample or exposing it to a dose of reactive gas.

Energy Range

Edge Energy Range

In the absorption edge region of metals, the photoelectron is excited to the first unoccupied level above the Fermi level. Therefore its mean free path in a pure single crystal at zero temperature is as large as infinite, and it remains very large, increasing the energy of the final state up to about 5 eV above the Fermi level. Beyond the role of the unoccupied density of states and matrix elements in single electron excitations, many-body effects appear as an "infrared singularity" at the absorption threshold in metals.

In the absorption edge region of insulators the photoelectron is excited to the first unoccupied level above the chemical potential but the unscreened core hole forms a localized bound state called core exciton.

EXAFS Energy Range

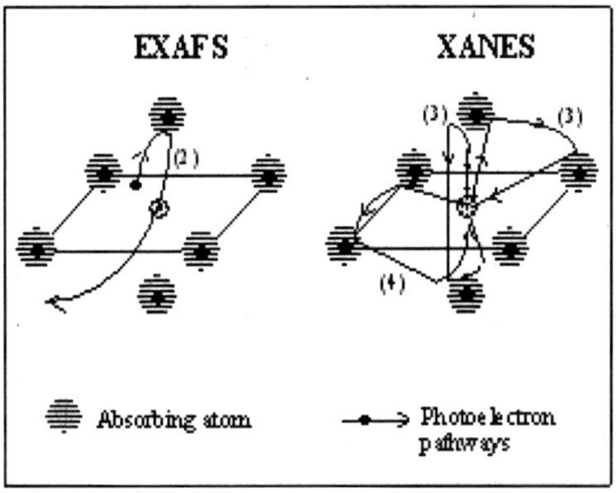

Figure: *Pictorial view of photoelectron scattering processes in the single-scattering regime, EXAFS (this assumes the single scattering approximation... multiple scattering can be considered with EXAFS), and in the multiple scattering regime, XANES. In EXAFS the photoelectron is scattered only by a single neighbour atom, in XANES all the scattering pathways, classified according to the number of scattering event (3), (4), (5) etc. contribute to the absorption cross section.*

The fine structure in the x-ray absorption spectra in the high energy range extending from about 150 eV beyond the ionization potential is a powerful tool to determine the atomic pair distribution (i.e. interatomic distances) with a time scale of about 10^{-15} s. In fact the final state of the excited photoelectron in the high kinetic energy range (150-2000 eV) is determined only by single backscattering events due to the low amplitude photoelectron scattering.

XANES Energy Range

In the XANES region, starting about 5 eV beyond the absorption threshold, because of the low kinetic energy range (5-150 eV) the photoelectron backscattering amplitude by neighbour atoms is very large so that multiple scattering events become dominant in the XANES spectra.

The different energy range between XANES and EXAFS can be also explained in a very simple manner by the comparison between the photoelectron wavelength λ and the interatomic distance of the photoabsorber-backscatterer pair. The photoelectron kinetic energy is connected with the wavelength λ by the following relation:

$$E_{kinetic} = h\nu - E_{binding} = \hbar^2 k^2 / (2m) = (2\pi)^2 \hbar^2 / (2m\lambda^2)$$

that means that for high energy the wavelength is shorter than interatomic distances and hence the EXAFS region corresponds to a single scattering regime; while for lower E, $\lambda$is larger than interatomic distances and the XANES region is associated with a multiple scattering regime.

Final States

The absorption peaks of XANES spectra are determined by multiple scattering resonances of the photoelectron excited at the atomic absorption site and scattered by neighbour atoms. The local character of the final states is determined by the short photoelectron mean free path, that is strongly reduced (down to about 0.3 nm at 50 eV) in this energy range because of inelastic scattering of the photoelectron by electron-hole excitations (excitons) and collective electronic oscillations of the valence electrons called plasmons.

Applications

The great power of XANES derives from its elemental specificity. Because the various elements have different core level energies, NEXAFS permits extraction of the signal from a surface monolayer or even a single buried layer in the presence of a huge background signal. Buried layers are very important in engineering applications, such as magnetic recording media buried beneath a surface lubricant or dopants below an electrode in an integrated circuit. Because XANES can also determine the chemical state of elements which are present in bulk in minute quantities, it has found widespread use in environmental chemistry and geochemistry. The ability of XANES to study buried atoms is due to its integration over all final states including inelastically scattered electrons, as opposed to photoemission and Auger spectroscopy, which study atoms only with a layer or two of the surface.

Much chemical information can be extracted from the XANES region: formal valence (very difficult to experimentally determine in a nondestructive way); coordination environment (e.g., octahedral, tetrahedral coordination) and subtle geometrical distortions of it.

Transitions to bound vacant states just above the Fermi level can be seen. Thus XANES spectra can be used as a probe of the unoccupied band structure of a material.

The near-edge structure is characteristic of an environment and valence state hence one of its more common uses is in fingerprinting: if you have a mixture of sites/compounds in a sample you can fit the measured spectra with a linear combinations of XANES spectra of known species and determine the proportion of each site/compound in the sample. One example of such a use is the determination of the oxidation state of the plutonium in the soil at Rocky Flats.

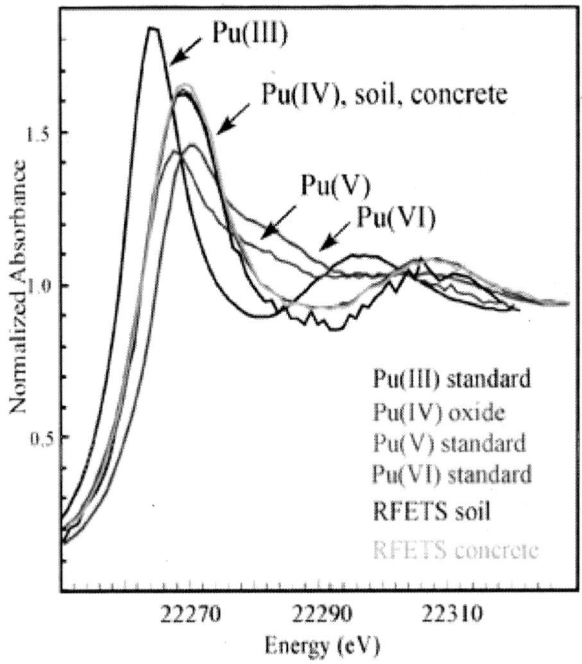

Figure: *The XANES experiments done on plutonium in soil, concrete and standards of the different oxidation states.*

SEXAFS

Surface-extended X-ray absorption fine structure (SEXAFS) is the surface-sensitive equivalent of the EXAFS technique. This technique involves the illumination of the sample by hig-intensity X-ray beams from a synchrotron and monitoring their photoabsorption by detecting in the intensity of Auger electrons as a function of the incident photon energy. Surface sensitivity is achieved by the interpretation of data depending on the intensity of the Auger electrons (which have an escape depth of ~1–2 nm) instead of looking at the relative absorption of the X-rays as in the parent method, EXAFS.

The photon energies are tuned through the characteristic energy for the onset of core level excitation for surface atoms. The core holes thus created can then be filled by nonradiative decay of a higher-lying electron and communication of energy to yet another electron, which can then escape from the surface (Auger emission). The photoabsorption can therefore be monitored by direct detection of these Auger electrons to the total photoelectron yield. The absorption coefficient versus incident photon energy contains oscillations which are due to the interference of the backscattered Auger electrons with the outward propagating waves. The period of this oscillations depends on the type of the backscattering atom and its distance from the central atom. Thus, this technique enables the investigation of interatomic distances for adsorbates and their coordination chemistry.

This technique benefits from long range order not being required, which sometimes becomes a limitation in the other conventional techniques like LEED (about 10 nm). This method also largely eliminates the background from the signal. It also benefits because it can probe different species in the sample by just tuning the X-ray photon energy to the absorption edge of that species. Joachim Stöhr played a major role in the initial development of this technique.

Experimental Setup

Synchrotron Radiation Sources

Normally, the SEXAFS work is done using synchrotron radiation as it has highly collimated, plane-polarized and precisely pulsed X-ray sources, with fluxes of 10^{12} to 10^{14} photons/sec/mrad/mA and greatly improves the signal-to-noise ratio over than obtainable from conventional sources. The experimental setup for the conventional EXAFS is shown here in Figure earlier. A bright source X-ray source is illuminating the sample and the transmission is being measured as the absorption coefficient as

$$\mu = \frac{\ln(I)}{\ln(I_o)},$$

where I is the transmitted and I_o is the incident intensity of the X-rays. Then it is plotted against the energy of the incoming X-ray photon energy.

Electron Detectors

In SEXAFS, an electron detector and a high-vacuum chamber is required to calculate the Auger yields instead of the intensity of the transmitted X-ray waves. The detector can be either an energy analyzer, as in the case of Auger measurements, or an electron multiplier, as in the case of total or partial secondary electron yield. The energy analyzer gives rise to better resolution while the electron multiplier has larger solid angle acceptance.

Signal-to-Noise Ratio

The equation governing the signal-to-noise ratio is

$$\frac{S}{N} = \sqrt{\frac{(\frac{\Omega}{4\pi}\epsilon_n\mu_A)}{(1+\frac{I_b}{I_n}(\mu_T + n))}}\left(\frac{\delta\mu_A}{\mu_A} I_o^{1/2}\right),$$

where

- μ_A is the absorption coefficient;
- I_n is the nonradiative contribution in electron counts/sec;
- I_b is the background contribution in electron counts/sec;
- μ_A is the absorption by the SEXAFS-producing element;
- μ_T is the total absorption by all the elements;
- I_o is the incident intensity;
- n is the attenuation length;
- $\Omega/(4\pi)$ is the solid angle acceptance for the detector;
- ε_n is the nonradiative yield which is the probability that the electron will not decay radiatively and will actually get emitted as an Auger electron.

Physics

Basics

The absorption of an X-ray photon by the atom excites a core level electron, thus generating a core hole. This generates a spherical electron wave with the excited atom as the centre. The wave propagates outwards and get scattered off from the neighbouring atoms and is turned back towards the central ionized atom. The oscillatory component of the photoabsorption originates from the coupling of this reflected wave to the initial state via the dipole operator M_{fs} as in (1). The Fourier transform of the oscillations gives the information about the spacing of the neighbouring atoms and their chemical environment. This phase information is carried over to the oscillations in the Auger signal because the transition time in Auger emission is of the same order of magnitude as the average time for a photoelectron in the energy range of interest. Thus, with a proper choice of the absorption edge and characteristic Auger transition, measurement of the variation of the intensity in a particular Auger line as a function of incident photon energy would be a measure of the photoabsorption cross section.

This excitation also triggers various decay mechanisms. These can be of radiative (fluorescence) or nonradiative (Auger and Coster–Kronig) nature. The intensity ratio between the Auger electron and X-ray emissions depends on the atomic number Z. The yield of the Auger electrons decreases with increasing Z.

Theory of EXAFS

The cross section of photoabsorption is given by Fermi's golden rule, which, in the dipole approximation, is given as

$$P = \frac{2\pi}{\hbar}\sum_f |M_{fs}|^2\,\delta(E_i + \hbar\omega - E_f),$$

$$M_{fs} = \langle f\,|\,e\epsilon\cdot\mathrm{r}\,|\,i\rangle,$$

where the initial state, i with energy E_i, consists of the atomic core and the Fermi sea, and the incident radiation field, the final state, f with energy E_f (larger than the Fermi level), consists of a core hole and an excited electron. ε is the polarization vector of the electric field, e the electron charge, and 'ω the x-ray photon energy. The photoabsorption signal contains a peak when the core level excitation is neared. It is followed by an oscillatory component which originates from the coupling of that part of the electron wave which upon scattering by the medium is turned back towards the central ionized atom, where it couples to the initial state via the dipole operator, M_i.

Assuming single-scattering and small-atom approximation for $kR_j >> 1$, where R_j is the distance from the central excited atom to the jth shell of neighbours and k is the photoelectrons wave vector,

$$k = \frac{1}{\hbar}\sqrt{[2m(\hbar(\omega - \omega_T) + V_o)]},$$

where 'ω_T is the absorption edge energy and V_o is the inner potential of the solid associated with exchange and correlation, the following expression for the oscillatory component of the photoabsorption cross section (for K-shell excitation) is obtained:

$$\chi(k) = k^{-1}\,|\,f(k,\pi)\,|\sum_j W_j \sin[2kR_j + \alpha(k)]\exp(-\gamma R_j - 2\sigma_j^2 k^2),$$

where the atomic scattering factor in a partial wave expansion with partial wave phase-shifts δ_l is given by

$$f(k,\theta) = (1/k)\sum_{l=0}^{\infty}(2l+1)[\exp(2i\delta_l(k)) - 1]P_l(\cos\theta).$$

$P_l(x)$ is the lth Legendre polynomial, γ is an attenuation coefficient, exp(–$2\sigma_i^2k^2$) is a Debye–Waller factor and weight W_j is given in terms of the number of atoms in the jth shell and their distance as

$$W_j = \frac{N_j}{R_j^2}.$$

The above equation for the $\chi(k)$ forms the basis of a direct, Fourier transform, method of analysis which has been successfully applied to the analysis of the EXAFS data.

Incorporation of EXAFS-Auger

The number of electrons arriving at the detector with an energy of the characteristic $W_\alpha XY$ Auger line (where W_α is the absorption edge core-level of element α, to which the incident x-ray line has been tuned) can be written as

$$N_T = N_{W_\alpha XY}(\hbar\omega) + N_B(\hbar\omega),$$

where N_B('ω) is the background signal and $N_{W_\alpha XY}(\hbar\omega)$ is the Auger signal we are interested in, where

$$N_{W_\alpha XY}(\hbar\omega) = (4\pi)^{-1}\psi_{W_\alpha XY}[1-\kappa]\int_\Omega \int_0^\infty \rho_\alpha(z)\, P_{W_\alpha}(\hbar\omega; z)\exp\left[\frac{-z}{\lambda(W_\alpha XY)}\cos\theta\right] dz d\Omega,$$

where $\psi_{W_\alpha XY}$ is the probability that an excited atom will decay via $W_\alpha XY$ Auger transition, $\rho_\alpha(z)$ is the atomic concentration of the element α at depth z, $\lambda(W_\alpha XY)$ is the mean free path for an $W_\alpha XY$ Auger electron, θ is the angle that the escaping Auger electron makes with the surface normal and κ is the photon emission probability which is dictated the atomic number. As the photoabsorption probability, $P_{W_\alpha}(\hbar\omega; z)$ is the only term that is dependent on the photon energy, the oscillations in it as a function of energy would give rise to similar oscillations in the $N_{W_\alpha XY}(\hbar\omega)$.

Electron Transport Chain

An electron transport chain (ETC) is a series of compounds that transfer electrons from electron donors to electron acceptors via redox reactions, and couples this electron transfer with the transfer of protons (H^+ ions) across a membrane. This creates an electrochemical proton gradient that drives ATP synthesis, or the generation chemical energy in the form of adenosine triphosphate (ATP).

Electron transport chains are used for extracting energy via redox reactions from sunlight in photosynthesis or, such as in the case of the oxidation of sugars, cellular respiration. In eukaryotes, the electron transport chain can be found in the inner mitochondrial membrane where it serves as the site of oxidative phosphorylation with ATP synthase. It is also found in the thylakoid membrane of the chloroplast in photosynthetic eukaryotes. In bacteria, the electron transport chain is located in their cell membrane.

In chloroplasts, light drives the conversion of water to oxygen and $NADP^+$ to NADPH with transfer of H^+ ions across chloroplast membranes. In mitochondria, it is the conversion of oxygen to water, NADH to NAD^+ and succinate to fumarate that are required to generate the proton gradient. Electron transport chains are major sites of premature electron leakage to oxygen, generating superoxide and potentially resulting in increased oxidative stress.

The electron transport chain consists of a spatially separated series of redox reactions in which electrons are transferred from a donor molecule to an acceptor molecule. The underlying force driving these reactions is the Gibbs free energy of the reactants and products. The Gibbs free energy is the energy available ("free") to do work. Any reaction that decreases the overall Gibbs free energy of a system is thermodynamically spontaneous.

The function of the electron transport chain is to produce a transmembrane proton electrochemical gradient as a result of the redox reactions. If protons flow back through the membrane, they enable mechanical work, such as rotating bacterial flagella. ATP synthase, an enzyme highly conserved among all domains of life, converts this mechanical work into chemical energy by producing ATP, which powers most cellular reactions. A small amount of ATP is available from substrate-level phosphorylation, for example, in glycolysis. In most organisms the majority of ATP is generated in electron transport chains, while only some obtain ATP by fermentation.

Electron Transport Chains in Mitochondria

Most eukaryotic cells have mitochondria, which produce ATP from products of the citric acid cycle, fatty acid oxidation, and amino acid oxidation. At the mitochondrial inner membrane, electrons from NADH and succinate pass through the electron transport chain to oxygen, which is reduced to water. The electron transport chain comprises an enzymatic series of electron donors and acceptors. Each electron donor passes electrons to a more electronegative acceptor, which in turn donates these electrons to another acceptor, a process that continues down the series until electrons are passed to oxygen, the most electronegative and terminal electron acceptor in the chain. Passage of electrons between donor and acceptor releases energy, which is used to generate a proton gradient across the mitochondrial membrane by actively "pumping" protons into the intermembrane space, producing a thermodynamic state that has the potential to do work. The entire process is called oxidative phosphorylation, since ADP is phosphorylated to ATP using the energy of hydrogen oxidation in many steps.

A small percentage of electrons do not complete the whole series and instead directly leak to oxygen, resulting in the formation of the free-radical superoxide, a highly reactive molecule that contributes to oxidative stress and has been implicated in a number of diseases and aging.

Mitochondrial Redox Carriers

Energy obtained through the transfer of electrons (blue arrows) down the ETC is used to pump protons (red arrows) from the mitochondrial matrix into the intermembrane space, creating an electrochemical proton gradient across the mitochondrial inner membrane (IMM) called ΔØ. This electrochemical proton gradient allows ATP synthase (ATP-ase) to use the flow of H^+ through the enzyme back into the matrix to generate ATP from adenosine diphosphate (ADP) and inorganic phosphate. Complex I (NADH coenzyme Q reductase; labelled I) accepts electrons from the Krebs cycle electron carrier nicotinamide adenine dinucleotide (NADH), and passes them to coenzyme Q (ubiquinone; labelled Q), which also receives electrons from complex II (succinate dehydrogenase; labelled II). UQ passes electrons to complex III (cytochrome bc_1 complex; labelled III), which passes them to cytochrome *c* (cyt *c*). Cyt *c* passes electrons to Complex IV (cytochrome *c* oxidase; labelled IV), which uses the electrons and hydrogen ions to reduce molecular oxygen to water.

Four membrane-bound complexes have been identified in mitochondria. Each is an extremely complex transmembrane structure that is embedded in the inner membrane. Three of them are proton pumps. The structures are electrically connected by lipid-soluble electron carriers and water-soluble electron carriers. The overall electron transport chain:

NADH → *Complex I* → Q → *Complex III* → cytochrome *c* → *Complex IV* → O_2

↑

Complex II

↑

FADH

Complex I

In *Complex I* (NADH dehydrogenase, also called NADH: ubiquinone oxidoreductase; EC 1.6.5.3) two electrons are removed from NADH and transferred to a lipid-soluble carrier, *ubiquinone* (Q). The reduced product, *ubiquinol* (QH_2) freely diffuses within the membrane, and Complex I translocates four protons (H^+) across the membrane, thus producing a proton gradient. Complex I is one of the main sites at which premature electron leakage to oxygen occurs, thus being one of the main sites of production of superoxide.

The Pathway of Electrons Occurs as Follows: NADH is oxidized to NAD^+, by reducing Flavin mononucleotide to $FMNH_2$ in one two-electron step. $FMNH_2$ is then oxidized in two one-electron steps, through a semiquinone intermediate. Each electron thus transfers from the $FMNH_2$ to an Fe-S cluster, from the Fe-S cluster to ubiquinone (Q). Transfer of the first electron results in the free-radical (semiquinone) form of Q, and transfer of the second electron reduces the semiquinone form to the ubiquinol form, QH_2. During this process, four protons are translocated from the mitochondrial matrix to the intermembrane space.

Complex II

In *Complex II* (succinate dehydrogenase; EC 1.3.5.1) additional electrons are delivered into the quinone pool (Q) originating from succinate and transferred (via FAD) to Q. Complex II consists of four protein subunits: SDHA, SDHB, SDHC, and SDHD. Other electron donors (e.g., fatty acids an glycerol 3-phosphate) also direct electrons into Q (via FAD). Complex 2 is a parallel electron transport pathway to complex 1, but unlike complex 1, no protons are transported to the intermembrane space in this pathway. Therefore, the pathway through complex 2 contributes less energy to the overall electron transport chain process.

Complex III

In *Complex III* (cytochrome *bc₁* complex; EC 1.10.2.2), the Q-cycle contributes to the proton gradient by an asymmetric absorption/release of protons. Two electrons are removed from QH_2 at the Q_O site and sequentially transferred to two molecules of cytochrome *c*, a water-soluble electron carrier located within the intermembrane space. The two other electrons sequentially pass across the protein to the Q_i site where the quinone part of ubiquinone is reduced to quinol. A proton gradient is formed by one quinol (2H+2e-) oxidations at the Q_o site to form one quinol (2H+2e-) at the Q_i site. (in total six protons are translocated: two protons reduce quinone to quinol and two protons are released from two ubiquinol molecules).

QH2 + 2 cytochrome c (FeIII) + 2 H+in → Q + 2 cytochrome c (FeII) + 4 H+out

When electron transfer is reduced (by a high membrane potential or respiratory inhibitors such as antimycin A), Complex III may leak electrons to molecular oxygen, resulting in superoxide formation.

Complex IV

In *Complex IV* (cytochrome *c* oxidase; EC 1.9.3.1), sometimes called cytochrome A3, four electrons are removed from four molecules of cytochrome *c* and transferred to molecular oxygen (O_2), producing two molecules of water. At the same time, four protons are removed from the mitochondrial matrix (although only two are translocated across the membrane), contributing to the proton gradient. The activity of cytochrome c oxidase is inhibited by cyanide.

Coupling with Oxidative Phosphorylation

According to the chemiosmotic coupling hypothesis, proposed by Nobel Prize in Chemistry winner Peter D. Mitchell, the electron transport chain and oxidative phosphorylation are coupled by a proton gradient across the inner mitochondrial membrane. The efflux of protons from the mitochondrial matrix creates an electrochemical gradient (proton gradient). This gradient is used by the F_0F_1 ATP synthase complex to make ATP via oxidative phosphorylation. ATP synthase is sometimes described as *Complex V* of the electron transport chain. The F_0 component of ATP synthase acts as an ion channel that provides for a proton flux back into the mitochondrial matrix. This reflux releases free energy produced during the generation of the oxidized forms of the electron carriers (NAD^+ and Q). The free energy is used to drive ATP synthesis, catalyzed by the F_1 component of the complex.

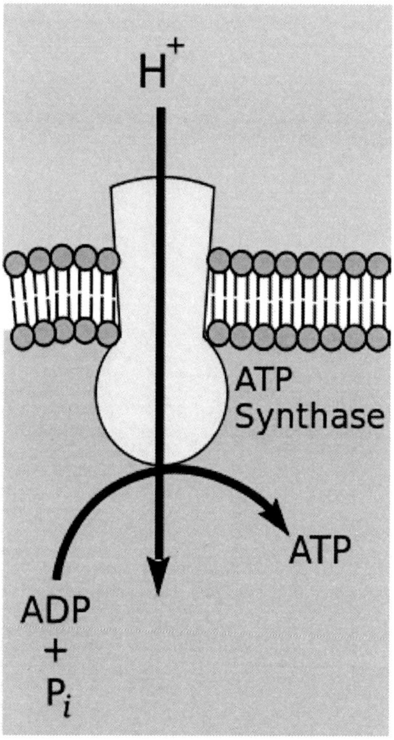

Figure: *Depiction of ATP synthase, the site of oxidative phosphorylation to generate ATP.*

Coupling with oxidative phosphorylation is a key step for ATP production. However, in specific cases, uncoupling the two processes may be biologically useful. The uncoupling protein, thermogenin—present in the inner mitochondrial membrane of brown adipose tissue—provides for an alternative flow of protons back to the inner mitochondrial matrix. This alternative flow results in thermogenesis rather than ATP production. Synthetic uncouplers (e.g., 2,4-dinitrophenol) also exist, and, at high doses, are lethal.

Summary

In the mitochondrial electron transport chain electrons move from an electron donor (NADH or QH_2) to a terminal electron acceptor (O_2) via a series of redox reactions. These reactions are coupled to the creation of a proton gradient across the mitochondrial inner membrane. There are three proton pumps: *I*, *III*, and *IV*. The resulting transmembrane proton gradient is used to make ATP via ATP synthase.

The reactions catalyzed by *Complex I* and *Complex III* work roughly at equilibrium. This means that these reactions are readily reversible, by increasing the concentration of the products relative to the concentration of the reactants (for example, by increasing the proton gradient). ATP synthase is also readily reversible. Thus ATP can be used to build a proton gradient, which in turn can be used to make NADH. This process of *reverse electron transport* is important in many prokaryotic electron transport chains.

Electron Transport Chains in Bacteria

In eukaryotes, NADH is the most important electron donor. The associated electron transport chain is

NADH → *Complex I* → Q → *Complex III* → cytochrome *c* → *Complex IV* → O_2 where *Complexes I, III* and *IV* are proton pumps, while Q and cytochrome *c* are mobile electron carriers. The electron acceptor is molecular oxygen.

In prokaryotes (bacteria and archaea) the situation is more complicated, because there are several different electron donors and several different electron acceptors. The generalized electron transport chain in bacteria is:

```
Donor          Donor                   Donor
                 ↓                       ↓                         ↓
               dehydrogenase  →       quinone   →   bc1   →   cytochrome
                                         ↓                         ↓
                            oxidase(reductase)          oxidase(reductase)
                                         ↓                         ↓
                                     Acceptor                  Acceptor
```

Note that electrons can enter the chain at three levels: at the level of a dehydrogenase, at the level of the quinone pool, or at the level of a mobile cytochrome electron carrier. These levels correspond to successively more positive redox potentials, or to successively decreased potential differences relative to the terminal electron acceptor. In other words, they correspond to successively smaller Gibbs free energy changes for the overall redox reaction *Donor* → *Acceptor*.

Individual bacteria use multiple electron transport chains, often simultaneously. Bacteria can use a number of different electron donors, a number of different dehydrogenases, a number of different oxidases and reductases, and a number of different electron acceptors. For example, *E. coli* (when growing aerobically using glucose as an energy source) uses two different NADH dehydrogenases and two different quinol oxidases, for a total of four different electron transport chains operating simultaneously.

A common feature of all electron transport chains is the presence of a proton pump to create a transmembrane proton gradient. Bacterial electron transport chains may contain as many as three proton pumps, like mitochondria, or they may contain only one or two. They always contain at least one proton pump.

Electron Donors

In the present day biosphere, the most common electron donors are organic molecules. Organisms that use organic molecules as an energy source are called *organotrophs*. Organotrophs (animals, fungi, protists) and *phototrophs* (plants and algae) constitute the vast majority of all familiar life forms.

Some prokaryotes can use inorganic matter as an energy source. Such organisms are called *lithotrophs* ("rock-eaters"). Inorganic electron donors include hydrogen, carbon monoxide, ammonia, nitrite, sulfur, sulfide, and ferrous iron. Lithotrophs have been found growing in rock formations thousands of metres below the surface of Earth. Because of their volume of distribution, lithotrophs may actually outnumber organotrophs and phototrophs in our biosphere.

The use of inorganic electron donors as an energy source is of particular interest in the study of evolution. This type of metabolism must logically have preceded the use of organic molecules as an energy source.

Dehydrogenases

Bacteria can use a number of different electron donors. When organic matter is the energy source, the donor may be NADH or succinate, in which case electrons enter the electron transport chain via NADH dehydrogenase (similar to *Complex I* in mitochondria) or succinate dehydrogenase (similar to *Complex II*). Other dehydrogenases may be used to process different energy sources: formate dehydrogenase, lactate dehydrogenase, glyceraldehyde-3-phosphate dehydrogenase, H_2 dehydrogenase (hydrogenase), etc. Some dehydrogenases are also proton pumps; others funnel electrons into the quinone pool. Most dehydrogenases show induced expression in the bacterial cell in response to metabolic needs triggered by the environment in which the cells grow.

Quinone Carriers

Quinones are mobile, lipid-soluble carriers that shuttle electrons (and protons) between large, relatively immobile macromolecular complexes embedded in the membrane. Bacteria use *ubiquinone* (the same quinone that mitochondria use) and related quinones such as *menaquinone*. Another name for ubiquinone is *Coenzyme Q10.*

Proton Pumps

A *proton pump* is any process that creates a proton gradient across a membrane. Protons can be physically moved across a membrane; this is seen in mitochondrial *Complexes I* and *IV*. The same effect can be produced by moving electrons in the opposite direction. The result is the disappearance of a proton from the cytoplasm and the appearance of a proton in the periplasm. Mitochondrial *Complex III* uses this second type of proton pump, which is mediated by a quinone (the Q cycle).

Some dehydrogenases are proton pumps; others are not. Most oxidases and reductases are proton pumps, but some are not. Cytochrome bc_1 is a proton pump found in many, but not all, bacteria (it is not found in *E. coli*). As the name implies, bacterial bc_1 is similar to mitochondrial bc_1 (*Complex III*).

Proton pumps are the heart of the electron transport process. They produce the transmembrane electrochemical gradient that enables ATP Synthase to synthesize ATP.

Cytochrome Electron Carriers

Cytochromes are pigments that contain iron. They are found in two very different environments.

Some cytochromes are water-soluble carriers that shuttle electrons to and from large, immobile macromolecular structures imbedded in the membrane. The mobile cytochrome electron carrier in mitochondria is cytochrome *c*. Bacteria use a number of different mobile cytochrome electron carriers.

Other cytochromes are found within macromolecules such as *Complex III* and *Complex IV*. They also function as electron carriers, but in a very different, intramolecular, solid-state environment.

Electrons may enter an electron transport chain at the level of a mobile cytochrome or quinone carrier. For example, electrons from inorganic electron donors (nitrite, ferrous iron, etc.) enter the electron transport chain at the cytochrome level. When electrons enter at a redox level greater than NADH, the electron transport chain must operate in reverse to produce this necessary, higher-energy molecule.

Terminal Oxidases and Reductases

When bacteria grow in aerobic environments, the terminal electron acceptor (O_2) is reduced to water by an enzyme called an *oxidase*. When bacteria grow in anaerobic environments, the terminal electron acceptor is reduced by an enzyme called a *reductase*.

In mitochondria the terminal membrane complex (*Complex IV*) is cytochrome oxidase. Aerobic bacteria use a number of different terminal oxidases. For example, *E. coli* does not have a cytochrome oxidase or a bc_1 complex. Under aerobic conditions, it uses two different terminal quinol oxidases (both proton pumps) to reduce oxygen to water.

Anaerobic bacteria, which do not use oxygen as a terminal electron acceptor, have terminal reductases individualized to their terminal acceptor. For example, *E. coli* can use fumarate reductase, nitrate reductase, nitrite reductase, DMSO reductase, or trimethylamine-N-oxide reductase, depending on the availability of these acceptors in the environment.

Most terminal oxidases and reductases are *inducible.* They are synthesized by the organism as needed, in response to specific environmental conditions.

Electron Acceptors

Just as there are a number of different electron donors (organic matter in organotrophs, inorganic matter in lithotrophs), there are a number of different electron acceptors, both organic and inorganic. If oxygen is available, it is invariably used as the terminal electron acceptor, because it generates the greatest Gibbs free energy change and produces the most energy.

In anaerobic environments, different electron acceptors are used, including nitrate, nitrite, ferric iron, sulfate, carbon dioxide, and small organic molecules such as fumarate.

Since electron transport chains are redox processes, they can be described as the sum of two redox pairs. For example, the mitochondrial electron transport chain can be described as the sum of the NAD^+/NADH redox pair and the O_2/H_2O redox pair. NADH is the electron donor and O_2 is the electron acceptor.

Not every donor-acceptor combination is thermodynamically possible. The redox potential of the acceptor must be more positive than the redox potential of the donor. Furthermore, actual environmental conditions may be far different from *standard* conditions (1 molar concentrations, 1 atm partial pressures, pH = 7), which apply to *standard* redox potentials. For example, hydrogen-evolving bacteria grow at an ambient partial pressure of hydrogen gas of 10^{-4} atm. The associated redox reaction, which is thermodynamically favourable in nature, is thermodynamically impossible under "standard" conditions.

Summary

Bacterial electron transport pathways are, in general, inducible. Depending on their environment, bacteria can synthesize different transmembrane complexes and produce different electron transport chains in their cell membranes. Bacteria select their electron transport chains from a DNA library containing multiple possible dehydrogenases, terminal oxidases and terminal reductases. The situation is often summarized by saying that electron transport chains in bacteria are *branched*, *modular*, and *inducible.*

Photosynthetic Electron Transport Chains

In oxidative phosphorylation, electrons are transferred from a low-energy electron donor (e.g., NADH) to an acceptor (e.g., O_2) through an electron transport chain. In photophosphorylation, the energy of sunlight is used to *create* a high-energy electron donor and an electron acceptor. Electrons are then transferred from the donor to the acceptor through another electron transport chain.

Photosynthetic electron transport chains have many similarities to the oxidative chains discussed above. They use mobile, lipid-soluble carriers (quinones) and mobile, water-soluble carriers (cytochromes, etc.). They also contain a proton pump. It is remarkable that the proton pump in *all* photosynthetic chains resembles mitochondrial *Complex III.*

Photosynthetic electron transport chains are discussed in greater detail in the articles Photophosphorylation, Photosynthesis, Photosynthetic reaction centre and Light-dependent reaction.

5

Scanning Tunnelling Microscope

A scanning tunnelling microscope (STM) is an instrument for imaging surfaces at the atomic level. Its development in 1981 earned its inventors, Gerd Binnig and Heinrich Rohrer (at IBMZurich), the Nobel Prize in Physics in 1986. For an STM, good resolution is considered to be 0.1 nm lateral resolution and 0.01 nm depth resolution. With this resolution, individual atoms within materials are routinely imaged and manipulated. The STM can be used not only in ultra-high vacuum but also in air, water, and various other liquid or gas ambients, and at temperatures ranging from near zero kelvin to a few hundred degrees Celsius.

The STM is based on the concept of quantum tunnelling. When a conducting tip is brought very near to the surface to be examined, a bias (voltage difference) applied between the two can allow electrons to tunnel through the vacuum between them. The resulting *tunnelling current* is a function of tip position, applied voltage, and the local density of states (LDOS) of the sample. Information is acquired by monitoring the current as the tip's position scans across the surface, and is usually displayed in image form. STM can be a challenging technique, as it requires extremely clean and stable surfaces, sharp tips, excellent vibration control, and sophisticated electronics.

Procedure

First, a voltage bias is applied and the tip is brought close to the sample by some coarse sample-to-tip control, which is turned off when the tip and sample are sufficiently close. At close range, fine control of the tip in all three dimensions when near the sample is typically piezoelectric, maintaining tip-sample separation W typically in the 4-7 A range, which is the equilibrium position between attractive (3<W<10A) and repulsive (W<3A) interactions. In this situation, the voltage bias will cause electrons to tunnel between the tip and sample, creating a current that can be measured. Once tunnelling is established, the tip's bias and position with respect to the sample can be varied (with the details of this variation depending on the experiment) and data are obtained from the resulting changes in current.

Fig. *A close-up of a simple scanning tunnelling microscope head using a platinum-iridium tip.*

If the tip is moved across the sample in the x-y plane, the changes in surface height and density of states cause changes in current. These changes are mapped in images. This change in current with respect to position can be measured itself, or the height, z, of the tip corresponding to a constant current can be measured. These two modes are called constant height mode and constant current mode, respectively. In constant current mode, feedback electronics adjust the height by a voltage to the piezoelectric height control mechanism. This leads to a height variation and thus the image comes from the tip topography across the sample and gives a constant charge density surface; this means contrast on the image is due to variations in charge density. In constant height mode, the voltage and height are both held constant while the current changes to keep the voltage from changing; this leads to an image made of current changes over the surface, which can be related to charge density. The benefit to using a constant height mode is that it is faster, as the piezoelectric movements require more time to register the height change in constant current mode, than the voltage change in constant height mode. All images produced by STM are grayscale, with colour optionally added in post-processing in order to visually emphasize important features.

In addition to scanning across the sample, information on the electronic structure at a given location in the sample can be obtained by sweeping voltage and measuring current at a specific location. This type of measurement is called scanning tunnelling spectroscopy (STS) and typically results in a plot of the local density of states as a function of energy within the sample. The advantage of STM over other measurements of the density of states lies in its ability to make extremely local measurements: for example, the density of states at animpurity site can be compared to the density of states far from impurities. Framerates of at least 1 Hz enable so called Video-STM (up to 50 Hz is possible). This can be used to scan surface diffusion.

Instrumentation

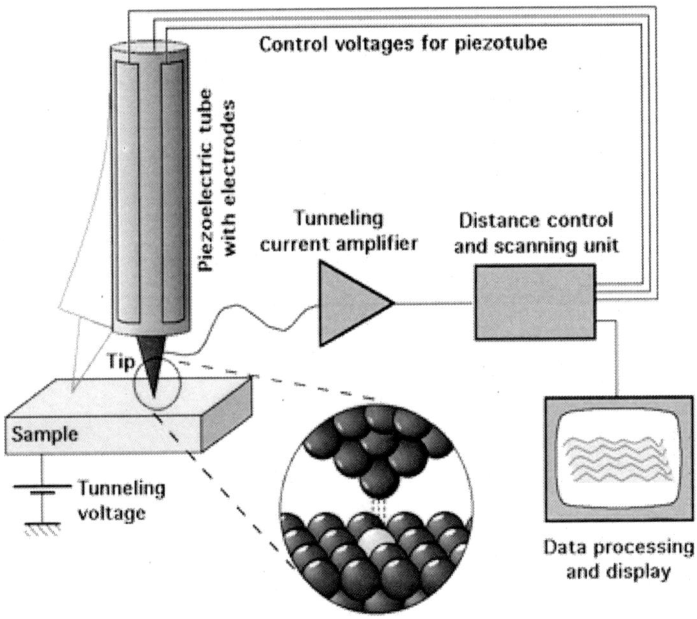

Fig. *Schematic view of an STM*

The components of an STM include scanning tip, piezoelectric controlled height and x,y scanner, coarse sample-to-tip control, vibration isolation system, and computer.

The resolution of an image is limited by the radius of curvature of the scanning tip of the STM. Additionally,image artifacts can occur if the tip has two tips at the end rather than a single atom; this leads to "double-tip imaging," a situation in which both tips contribute to the tunnelling. Therefore it has been essential to develop processes for consistently obtaining sharp, usable tips. Recently, carbon nanotubes have been used in this instance.

The tip is often made of tungsten or platinum-iridium, though gold is also used. Tungsten tips are usually made by electrochemical etching, and platinum-iridium tips by mechanical shearing.

Due to the extreme sensitivity of tunnel current to height, proper vibration isolation or an extremely rigid STM body is imperative for obtaining usable results. In the first STM by Binnig and Rohrer, magnetic levitation was used to keep the STM free from vibrations; now mechanical spring or gas spring systems are often used. Additionally, mechanisms for reducing eddy currents are sometimes implemented. Maintaining the tip position with respect to the sample, scanning the sample and acquiring the data is computer controlled. The computer may also be used for enhancing the image with the help of image processing as well as performing quantitative measurements.

Other STM Related Studies

Many other microscopy techniques have been developed based upon STM. These include photon scanning microscopy (PSTM), which uses an optical tip to tunnel photons; scanning tunnelling potentiometry (STP), which measures electric potential across a surface; spin polarised scanning tunnelling microscopy (SPSTM), which uses a ferromagnetic tip to tunnel spin-polarised electrons into a magnetic sample, and atomic force microscopy (AFM), in which the forcecaused by interaction between the tip and sample is measured.

Other STM methods involve manipulating the tip in order to change the topography of the sample. This is attractive for several reasons. Firstly the STM has an atomically precise positioning system which allows very accurate atomic scale manipulation. Furthermore, after the surface is modified by the tip, it is a simple matter to then image with the same tip, without changing the instrument. IBM researchers developed a way to manipulate Xenon atoms adsorbed on a nickel surface. This technique has been used to create electron "corrals" with a small number of adsorbed atoms, which allows the STM to be used to observe electron Friedel oscillations on the surface of the material. Aside from modifying the actual sample surface, one can also use the STM to tunnel electrons into a layer of E-Beam photoresist on a sample, in order to do lithography. This has the advantage of offering more control of the exposure than traditional Electron beam lithography. Another practical application of STM is atomic deposition of metals (Au, Ag, W, etc.) with any desired (pre-programmed) pattern, which can be used as contacts to nanodevices or as nanodevices themselves.

Recently groups have found they can use the STM tip to rotate individual bonds within single molecules. The electrical resistance of the molecule depends on the orientation of the bond, so the molecule effectively becomes a molecular switch.

Principle of Operation

Tunnelling is a functioning concept that arises from quantum mechanics. Classically, an object hitting an impenetrable barrier will not pass through. In contrast, objects with a very small mass, such as the electron, have wavelike characteristics which permit such an event, referred to as tunnelling.

Electrons behave as beams of energy, and in the presence of a potential $U(z)$, assuming 1-dimensional case, the energy levels $ø_n(z)$ of the electrons are given by solutions to Schrodinger's equation,

$$-\frac{\hbar^2}{2m}\frac{\partial^2\psi_n(z)}{\partial z^2}+U(z)\psi_n(z)=E\psi_n(z)$$

where ' is the reduced Planck's constant, z is the position, and m is the mass of an electron. If an electron of energy E is incident upon an energy barrier of height $U(z)$, the electron wave function is a traveling wave solution,

$$\psi_n(z)=\psi_n(0)e^{\pm ikz}$$

where

$$k=\frac{\sqrt{2m(E-U(z))}}{\hbar}$$

if $E > U(z)$, which is true for a wave function inside the tip or inside the sample. Inside a barrier, $E < U(z)$ so the wave functions which satisfy this are decaying waves,

where

$$\kappa=\frac{\sqrt{2m(U-E)}}{\hbar}$$

quantifies the decay of the wave inside the barrier, with the barrier in the $+z$ direction for " e.

Knowing the wave function allows one to calculate the probability density for that electron to be found at some location. In the case of tunnelling, the tip and sample wave functions overlap such that when under a bias, there is some finite probability to find the electron in the barrier region and even on the other side of the barrier. Let us assume the bias is V and the barrier width is W. This probability, P, that an electron at z=0 (left edge of barrier) can be found at z=W (right edge of barrier) is proportional to the wave function squared,

$$P \propto |\psi_n(0)|^2\, e^{-2\kappa W}.$$

If the bias is small, we can let U " E H" oM in the expression for e, where oM, the work function, gives the minimum energy needed to bring an electron from an occupied level, the highest of which is at the Fermilevel (for metals at T=0 kelvins), to vacuum level. When a small bias V is applied to the system, only electronic states very near the Fermilevel, within eV (a product of electron charge and voltage, not to be confused here with electronvolt unit), are excited. These excited electrons can tunnel across the barrier. In other words, tunnelling occurs mainly with electrons of energies near the Fermilevel.

However, tunnelling does require that there is an empty level of the same energy as the electron for the electron to tunnel into on the other side of the barrier. It is because of this restriction that the tunnelling current can be related to the density of available or filled states in the sample. The current due to an applied voltage V (assume tunnelling occurs sample to tip) depends on two factors: 1) the number of electrons between E_f and eV in the sample, and 2) the number among them which have corresponding free states to tunnel into on the other side of the barrier at the tip. The higher density of available states the greater the tunnelling current. When V is positive, electrons in the tip tunnel into empty states in the sample; for a negative bias, electrons tunnel out of occupied states in the sample into the tip.

Mathematically, this tunnelling current is given by

$$I \propto \sum_{E_f - eV}^{E_f} |\psi_n(0)|^2 e^{-2\kappa W}.$$

One can sum the probability over energies between E_f " eV and E_f to get the number of states available in this energy range per unit volume, thereby finding the local density of states (LDOS) near the Fermilevel. The LDOS near some energy E in an interval a is given by

$$\rho_s(z, E) = \frac{1}{\epsilon} \sum_{E-\epsilon}^{E} |\psi_n(z)|^2,$$

and the tunnel current at a small bias V is proportional to the LDOS near the Fermilevel, which gives important information about the sample. It is desirable to use LDOS to express the current because this value does not change as the volume changes, while probability density does. Thus the tunnelling current is given by

$$I \propto V \rho_s(0, E_f) e^{-2\kappa W}$$

where $n_s(0,E_f)$ is the LDOS near the Fermilevel of the sample at the sample surface. This current can also be expressed in terms of the LDOS near the Fermilevel of the sample at the tip surface,

$$I \propto V \rho_s(W, E_f)$$

The exponential term in the above equations means that small variations in W greatly influence the tunnel current. If the separation is decreased by 1 ú, the current increases by an order of magnitude, and vice versa.

This approach fails to account for the *rate* at which electrons can pass the barrier. This rate should affect the tunnel current, so it can be treated using the Fermi's golden rule with the appropriate tunnelling matrix element. John Bardeen solved this problem in his study of the metal-insulator-metal junction.

He found that if he solved Schrodinger's equation for each side of the junction separately to obtain the wave functions ø and ÷ for each electrode, he could obtain the tunnel matrix, M, from the overlap of these two wave functions. This can be applied to STM by making the electrodes the tip and sample, assigning ø and ÷ as sample and tip wave functions, respectively, and evaluating M at some surface S between the metal electrodes, where z=0 at the sample surface and z=W at the tip surface.

Now, Fermi's Golden Rule gives the rate for electron transfer across the barrier, and is written

$$w = \frac{2\pi}{\hbar} | M |^2 \delta(E_\psi - E_\chi),$$

where a($E_ø$–$E_÷$) restricts tunnelling to occur only between electron levels with the same energy. The tunnel matrix element, given by

$$M = \frac{\hbar^2}{2m} \int_{z=z_0} (\chi^* \frac{\partial \psi}{\partial z} - \psi \frac{\partial \chi^*}{\partial z}) dS,$$

is a description of the lower energy associated with the interaction of wave functions at the overlap, also called the resonance energy.

Summing over all the states gives the tunnelling current as

$$I = \frac{4\pi e}{\hbar} \int_{-\infty}^{+\infty} [E_f - eV + \epsilon) - f(E_f + \epsilon)](E_f - eV + \epsilon)\rho_T(E_f + \epsilon) | M |^2 \, de,$$

where f is the Fermifunction, n_s and n_T are the density of states in the sample and tip, respectively. The Fermidistribution function describes the filling of electron levels at a given temperature T.

Early Invention

An earlier, similar invention, the *Topografiner* of R. Young, J. Ward, and F. Scire from the NIST, relied on field emission. However, Young is credited by the Nobel Committee as the person who realised that it should be possible to achieve better resolution by using the tunnel effect.

Millipede Memory

Millipede is a non-volatile computer memory stored on nanoscopic pits burned into the surface of a thin polymer layer, read and written by a MEMS-based probe. It promises a data density of more than 1 terabit per square inch (1 gigabit per square millimeter), about 4 times the density of magnetic storage available today.

Millipede storage technology is being pursued as a potential replacement for magnetic recording in hard drives, at the same time reducing the form-factor to that of flash media. IBM demonstrated a prototype millipede storage device at CeBIT 2005, and was trying to make the technology commercially available by the end of 2007. At launch, it would probably be more expensive per-megabyte than prevailing technologies, but this disadvantage is hoped to be offset by the sheer storage capacity that Millipede technology would offer.

Technology—Basic Concept

The main memory of modern computers is constructed from one of a number of DRAM-related devices. DRAM basically consists of a series of capacitors, which store data as the presence or absence of electrical charge. Each capacitor and its associated control circuitry, referred to as a *cell*, holds one bit, and bits can be read or written in large blocks at the same time.

In contrast, hard drives store data on a disk that is covered with a magnetic material; data is represented as local magnetisation of this material. Reading and writing are accomplished by a single "head", which waits for the requested memory location to pass under the head while the disk spins. As a result, the drive's performance is limited by the mechanical speed of the motor, and is generally hundreds of thousands of times slower than DRAM. However, since the "cells" in a hard drive are much smaller, the storage density is much higher than DRAM.

Millipede storage attempts to combine the best features of both. Like the hard drive, millipede stores data in a "dumb" medium that is simpler and smaller than any cell used in an electronic medium. It accesses the data by moving the medium under the "head" as well. However, millipede uses many nanoscopic heads that can read and write in parallel, thereby dramatically increasing the throughput to the point where it can compete with some forms of electronic memory. Additionally, millipede's physical medium stores a bit in a very small area, leading to densities even higher than current hard drives.

Mechanically, millipede uses numerous *atomic force probes*, each of which is responsible for reading and writing a large number of bits associated with it. Bits are stored as a pit, or the absence of one, in the surface of a thermo-active polymer deposited as a thin film on a carrier known as the *sled*.

Any one probe can only read or write a fairly small area of the sled available to it, a *storage field*. Normally the sled is moved to position the selected bits under the probe using electromechanical actuators similar to those that position the read/write head in a typical hard drive, although the actual distance moved is tiny. The sled is moved in a scanning pattern to bring the requested bits under the probe, a process known as *x/y scan*.

The amount of memory serviced by any one field/probe pair is fairly small, but so is its physical size. Many such field/probe pairs are used to make up a memory device. Data reads and writes can be spread across many fields in parallel, increasing the throughput and improving the access times. For instance, a single 32-bit value would normally be written as a set of single bits sent to 32 different fields. In the initial experimental devices, the probes were mounted in a 32x32 grid for a total of 1,024 probes. Their layout looked like the legs on a millipede, and the name stuck.

The design of the cantilever array is the trickiest part, as it involves making numerous mechanical cantilevers, on which a probe has to be mounted. All the cantilevers are made entirely out of silicon, using surface micromachining at the wafer surface.

Reading and Writing Data

Each probe in the cantilever array stores and reads data thermo-mechanically, handling one bit at a time.

To accomplish a read, the probe tip is heated to around 300 °C and moved in proximity to the data sled. If the probe is located over a pit the cantilever will push it into the hole, increasing the surface area in contact with the sled, and in turn increasing the cooling as heat leaks into the sled from the probe. In the case where there is no pit at that location, only the very tip of the probe remains in contact with the sled, and the heat leaks away more slowly. The electrical resistance of the probe is a function of its temperature, rising with increasing temperature. Thus when the probe drops into a pit and cools, this registers as a drop in resistance. A low resistance will be translated to a "1" bit, or a "0" bit otherwise. While reading an entire storage field, the tip is dragged over the entire surface and the resistance changes are constantly monitored.

To write a bit, the tip of the probe is heated to a temperature above the *glass transition temperature* of the polymer used to manufacture the data sled, which is generally acrylic glass. In this case the transition temperature is around 400 °C. To write a "1", the polymer in proximity to the tip is softened, and then the tip is gently touched to it, causing a dent. To erase the bit and return it to the zero state, the tip is instead pulled up from the surface, allowing surface tension to pull the surface flat again. Older experimental systems used a variety of erasure techniques that were generally more time consuming and less successful. These older systems offered around 100,000 erases, but the available references do not contain enough information to say if this has been improved with the newer techniques.

As one might expect, the need to heat the probes requires a fairly large amount of power for general operation. However, the exact amount is dependent on the speed that data is being accessed; at slower rates the cooling during read is smaller, as is the number of times the probe has to be heated to a higher temperature to write. When operated at data rates of a few megabits per second, Millipede is expected to consume about 100 milliwatts, which is in the range of flash memory technology and considerably below hard drives. However, one of the main advantages of the Millipede design is that it is highly parallel, allowing it to run at much higher speeds into the GB/s. At these sorts of speeds one might expect power requirements more closely matching current hard drives.

Data transfer speed is limited to the kilobits-per-second range for an individual probe, which amounts to a few megabits for an entire array. Experiments done at IBM's Almaden Research Centre showed that individual tips could support data rates as high as 1 - 2 megabits per second, potentially offering aggregate speeds in the GB/s range.

Current State of the Art

The progress of millipede storage to a commercially useful product has been slower than expected. Huge advances in other competing storage systems, notably Flash and hard drives, has made the existing demonstrators unattractive for commercial production. Millipede appears to be in a race, attempting to mature quickly enough at a given technology level that it has not been surpassed by newer generations of the existing technologies by the time it is ready for production. The earliest generation millipede devices used probes 10 nanometers in diameter and 70 nanometers in length, producing pits about 40 nm in diameter on fields 92 μm x 92 μm. Arranged in a 32 x 32 grid, the resulting 3 mm x 3 mm chip stores 500 megabits of data or 62.5 MB, resulting in an areal density, the number of bits per square inch, on the order of 200 Gbit/in^2. IBM initially demonstrated this device in 2003, planning to introduce it commercially in 2005. By that point hard drives were approaching 150 Gbit/in^2, and have since surpassed it.

More recent devices demonstrated at CeBIT in 2005 have improved on the basic design, using a 64 x 64 cantilever chips with a 7 mm x 7 mm data sled, boosting the data storage capacity to 800 Gbit/in^2 using smaller pits. It appears the pit size can scale to about 10 nm, resulting in a theoretical areal density just over 1Tbit/in^2. IBM planned to introduce devices based on this sort of density in 2007. For comparison, perpendicular recording hard drives featured areal densities on the order of 329 Gbit/in^2 at the beginning of 2009, and has been predicted to top out at about 1 Tbit/in^2. Flash reached almost 250 Gbit/in^2 in early 2010.

Usage Scenarios

Micro Drives: Millipede systems can be used for micro drives, which will feature very small form factor, enabling use in small footprint devices like watches, mobile phones and personal media systems, and at the same time provide high capacity. The very high data density of millipede systems makes them a very good candidate to be put to this use.

High-capacity Hard Drives

The Millipede system provides high data density, low seek times, low power consumption and, probably, high reliability. These features make them candidates for building high capacity hard drives, with storage capacity in the range of terabytes. Although the data density of a Millipede is high, the capacity of an individual device is expected to be relatively low—on the order of single gigabytes. Thus replacing hard drives probably requires economically collecting around 100 Millipede devices into a single enclosure.

Sarfus

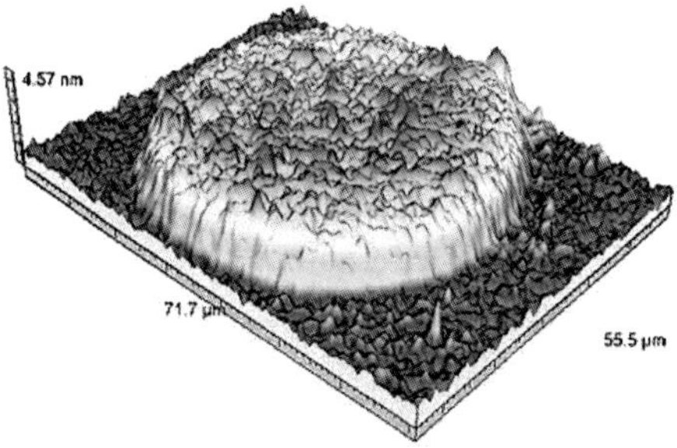

Fig. *3D Sarfus image of a DNA biochip.*

Sarfus is an optical quantitative imaging technique based on the association of:

- an upright or inverted optical microscope in crossed polarisation configuration and
- specific supporting plates - called surfs - on which the sample to observe is deposited.

Sarfus visualisation is based on the perfect control of the reflection properties of polarised light on a surf, which leads to increase the axial sensitivity of optical microscope by a factor of around 100 without reducing its lateral resolution. Thus this new technique increases the sensitivity of standard optical microscope to a point that it becomes possible to directly visualise thin films (down to 0.3 nanometer) and isolated nano-objects in real-time but also in air or in water.

Principles

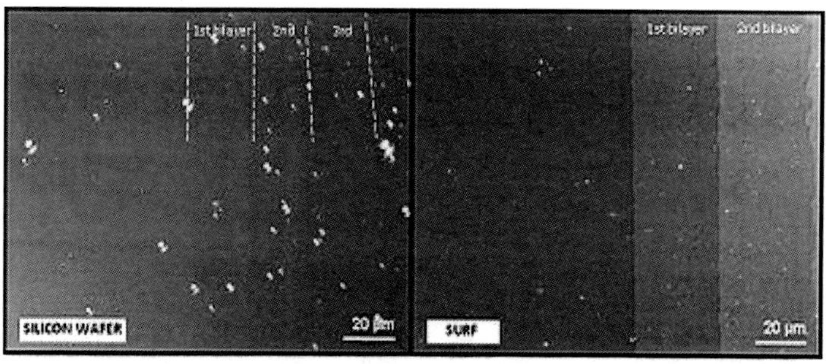

Fig. *Observation with standard optical microscope between cross polarisers of Langmuir-Blodgett layers (bilayer thickness: 5.4 nm) on silicon wafer and on surf*

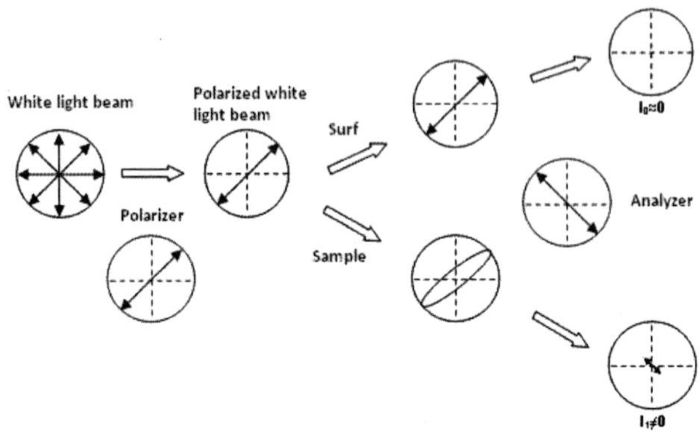

Fig. *Light polarisation after reflexion on a surf (0) and on nanoscale sample on a surf (1).*

A recent study on polarised light coherence leads to the development of new supports – the surfs – having contrast amplification properties for standard optical microscopy in cross polarisers mode. Made of optical layers on an opaque or transparent substrate, these supports do not modify the light polarisation after reflection even if the numerical aperture of the incident source is important. This property is modified when a sample is present on a surf, a non-null light component is then detected after the analyser rendering the sample visible.

The performances of these supports are estimated from the measurement of the contrast (C) of the sample defines by: $C = (I_1\text{-}I_0)/(I_0+I_1)$ where I_0 and I_1 represent the intensities reflected by the bare surf and by the analysed sample on the surf, respectively. For a one nanometer-film thickness, the surfs display a contrast 200 times higher than on silicon wafer. This high contrast increase allows the visualisation with standard optical microscope of films with thicknesses down to 0.3 nm, as well as nano-objects (down to 2 nm diameter) and this, without any kind of sample labelling (neither fluorescence, nor radioactive marker). An illustration of the contrast enhance is given hereafter with the observation in optical microscopy between cross polarisers of a Langmuir-Blodgett structure on a silicon wafer and on a surf.

In addition to visualisation, recent developments have allowed accessing to the thickness measurement of the analysed sample. A colorimetric correspondence is carried out between a calibration standard made of nano-steps and the analysed sample. Indeed, due to optical interference, a correlation exists between RGB (red, green, blue) parameters of the sample and its optical thickness. This leads to 3D-representation of the analysed samples, the measurement of profile sections, roughness and other topological measurements.

Experimental Setup

The experimental set-up is simple: the sample to be characterised is deposited by usual deposit techniques such as dip-coating, spin-coating, deposit pipette, evaporation... on a surf instead of the traditional microscope slide. The support is then placed on the microscope stage.

Synergy with Existing Equipments

The sarfus technique can be integrated in existing analysis equipment (Atomic force microscope (AFM), Raman spectroscopy, etc.) to add new functionalities, such as optical image, thickness measurement, kinetic study, and also for sample pre-localisation to save time and consumables (AFM tips, etc.).

Applications

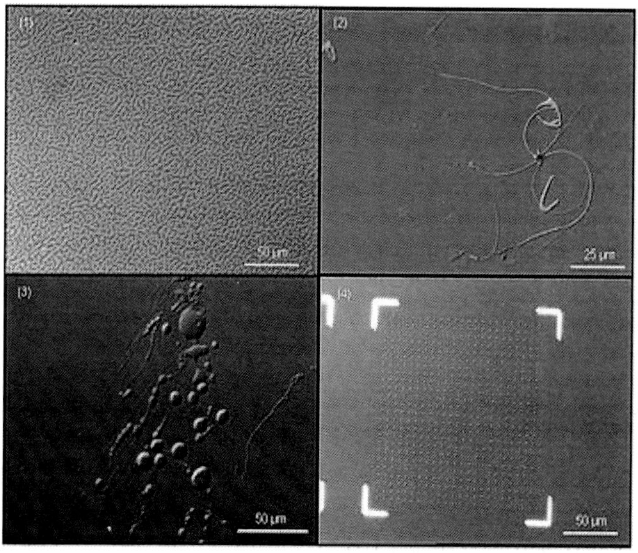

Fig. *Sarfus images of nanostructures: 1. Copolymer film microstructuration (73 nm), 2. Carbon nanotube bundles, 3. Lipid vesicles in aqueous solutions, 4. Nanopatterning of gold dots (50 nm^3).*

Life Sciences

- Biological films
- Biochip
- Phospholipids
- Soft lithography
- Cell adhesion.

Thin Films and Surface Treatment

- Polymers films
- Langmuir-Blodgett films
- Liquid crystals

- Plasma treatment
- Self-assembled monolayers

Nanomaterials

- Carbon nanotubes
- Nanoparticles
- Nanowires
- Graphene
- DNA molecules.

Advantages

Optical microscopy has several advantages compared to the usual techniques of nanocharacterisation. It is easy-to-use and directly vizualizes the sample. The analysis in real-time allows kinetic studies (real-time crystallisation, dewetting, etc.). The broad choice of magnification (2.5 to 100x) allows fields of view from several mm^2 to a few tens μm^2. Observations can be performed in controlled atmosphere and temperature.

Boride

In chemistry a boride is a chemical compound between boron and a less electronegative element, for example silicon boride (SiB_3 and SiB_6). The borides are a very large group of compounds that are generally high melting and are not ionic in nature. Some borides exhibit very useful physical properties. The term boride is also loosely applied to compounds such as $B_{12}As_2$ (N.B. Arsenic has an electronegativity higher than boron) that is often referred to as icosahedral boride.

Ranges of Compounds

The borides can be classified loosely as boron rich or metal rich, for example the compound YB_{66} at one extreme through to $Nd_2Fe_{14}B$ at the other. The generally accepted definition is that if the ratio of boron atoms to metal atoms is 4:1 or more the compound is boron rich, if it is less, then it is metal rich.

Boron Rich Borides (B:M 4:1 or More)

The main group metals, lanthanides and actinides form a wide variety of boron-rich borides, with metal:boron ratios up to YB_{66}.

The properties of this group vary from one compound to the next, and includes examples of compounds that are semiconductors, superconductors, diamagnetic, paramagnetic, ferromagnetic or anti-ferromagnetic. They are mostly stable and refractory.

Some metallic dodecaborides contain boron icosahedra, others (for example yttrium, zirconium and uranium) have the boron atoms arranged in cuboctahedra.

LaB_6 is an inert refractory compound, used in hot cathodes because of its low work function for emission of electrodes; YB_{66} crystals, grown by an indirect-heating floating zone method, are used as monochromators for low-energy synchrotron X-rays.

Metal Rich Borides (B:M Less Than 4:1)

The transition metals tend to form metal rich borides. Metal-rich borides, as a group, are inert and have high melting temperature. Some are easily formed and this explains their use in making turbine blades, rocket nozzles, etc. Some examples include AlB_2 and TiB_2. Recent investigations into this class of borides have revealed a wealth of interesting properties such as super conductivity at 39 K in MgB_2 and the ultra-incompressibility of OsB_2 and ReB_2.

Boride Structures

The boron rich borides contain 3-dimensional frameworks of boron atoms that can include boron polyhedra. The metal rich borides contain single boron atoms, B_2 units, boron chains or boron sheets/layers.

Examples of the different types of borides are:

- isolated boron atoms, example Mn_4B
- B_2 units, example V_3B
- chains of boron atoms, example FeB
- sheets or layers of boron atoms CrB_2
- 3-dimensional boron frameworks that include boron polyhedra, example NaB_{15} with boron icosahedra

Inorganic Nanotube

An inorganic nanotube is a cylindrical molecule often composed of metal oxides, and morphologically similar to a carbon nanotube. Inorganic nanotubes have been observed to occur naturally in some mineral deposits. Although Linus Pauling mentioned the possibility of curved layers in minerals as early as 1930, synthetic inorganic nanotubes did not appear until Reshef Tenne *et al.* reported the synthesis of nanotubes composed of tungsten disulfide (WS_2) in 1992.

In the intervening years, nanotubes have been synthesised of many other inorganic materials, such as vanadium oxide and manganese oxide, and are being researched for such applications asredox catalysts and cathode materials for batteries.

Fig. *Scanning electron microscopy image of* WS_2 *nanotube bundles.*

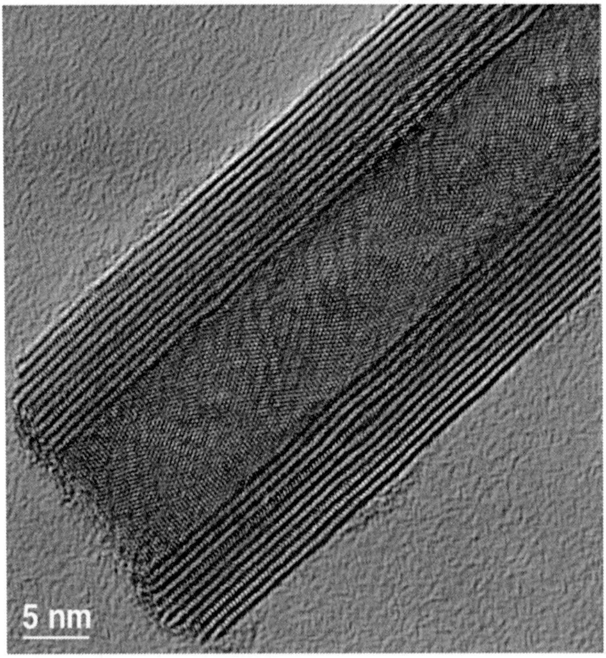

Fig. *Transmission electron microscopy image of an individual* WS_2 *multi-wall nanotube.*

History and Occurrence

Inorganic nanotubes are morphologically similar to carbon nanotubes and are observed in some mineral deposits of natural origin. Synthetic structures of this type were first reported by the group of Reshef Tennc in 1992.

Materials

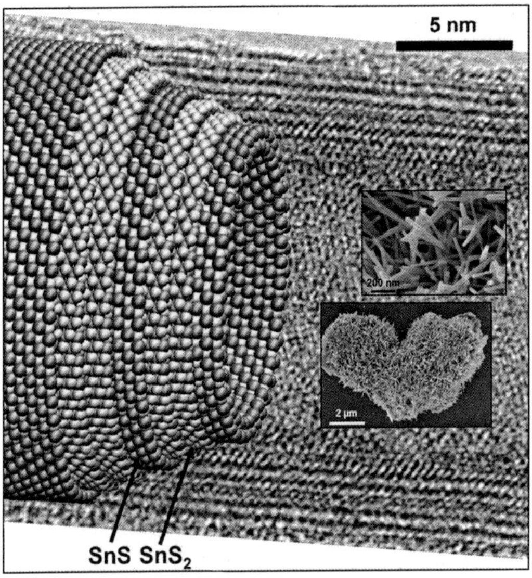

Fig. Illustration of SnS_2/SnS nanotubesuperstructures, made on basis of high-resolution transmission electron microscopy images.

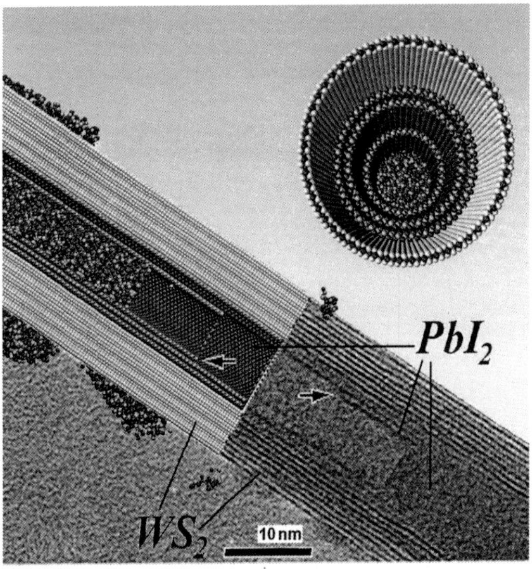

Fig. Illustration of PbI_2/WS_2 core-shell nanostructure.

Typical inorganic nanotube materials are 2D layered solids such as WS_2, MoS_2 and SnS_2, and WS_2 and SnS_2/SnS nanotubes have been synthesized in macroscopic amounts. However, traditional ceramics like titanium dioxide (TiO_2) and zinc oxide (ZnO) also form inorganic nanotubes. More recent nanotube and nanowire materials are transition metal/chalcogen/halogenides (TMCH), described by the formula $TM_6C_yH_z$, where TM is transition metal (molybdenum, tungsten, tantalum, niobium), C is chalcogen (sulfur, selenium, tellurium), H is halogen (iodine), and the composition is given by $8.2<(y+z)<10$. TMCH tubes can have a subnanometer-diameter, lengths tunable from hundreds of nanometers to tens of microns and show excellent dispersiveness owing to extremely weak mechanical coupling between the tubes.

In 2007, Chinese scientists announced the creation in the laboratory of copper and bismuthnanotubes.

Properties and Potential Applications

Inorganic nanotubes are an alternative material to better-explored carbon nanotubes, showing advantages such as easy synthetic access and high crystallinity, good uniformity and dispersion, predefined electrical conductivity depending on the composition of the starting material and needle-like morphology, good adhesion to a number of polymers and high impact-resistance. They are therefore promising candidates as fillers for polymer composites with enhanced thermal, mechanical, and electrical properties. Target applications for this kind of composites are materials for heat management, electrostatic dissipaters, wear protection materials, photovoltaic elements, etc. Inorganic nanotubes are heavier than carbon nanotubes and not as strong under tensile stress, but they are particularly strong under compression, leading to potential applications in impact-resistant applications such as bulletproof vests.

The mechanical strength of cellulose fibres can be increased by an order of magnitude by adding only 0.1 wt% of TMCH nanotubes, and measurements of electrical conductivity of polycaprolactone doped with TMCH nanotubes revealed a percolative behavior with an extremely low percolation threshold. The addition of WS_2 nanotubes to epoxy resin improved adhesion,f racture toughness and strain energy release rate. The wear of the nanotubes-reinforced epoxy was eight times lower than that of pure epoxy. WS_2 nanotubes were also embedded into apoly(methyl methacrylate) (PMMA) nanofibre matrix via electrospinning. The nanotubes were well dispersed and aligned along fiber axis. The enhanced stiffness and toughness of PMMA fiber meshes by means of inorganic nanotubes addition may have potential applications as impact-absorbing materials.

Optical properties of semiconductor quantum dot-inorganic nanotube hybrids reveal efficient resonant energy transfer from the quantum dot to the inorganic nanotubes upon photoexcitation. Nanodevices based on one-dimensional nanomaterials are thought for next-generation electronic and photoelectronic systems haveing small size, faster transport speed, higher efficiency and less energy consumption. A high-speed photodetector for visible and near-infrared light based on individual WS_2 nanotubes has been prepared in laboratory. Inorganic nanotubes are hollow and can be filled with another material, to preserve or guide it to a desired location or generate new properties in the filler material which is confined within a nanometer-scale diameter. To this goal, inorganic nanotube hybrids were made by filling WS_2 nanotubes with molten lead, antimony or bismuth iodide salt by a capillary wetting process, resulting in $PbI_2 \cong WS_2$, $SbI_3 \cong WS_2$ or $BiI_3 \cong WS_2$ core-shell nanotubes.

Hybrid Material

Hybrid materials are composites consisting of two constituents at the nanometer or molecular level. Commonly one of these compounds isinorganic and the other one organic in nature. Thus, they differ form traditional composites where the constituents are at the macroscopic (micrometer to millimeter level). Mixing at the microscopic scale leads to a more homogeneous material that either show characteristics in between the two original phases or even new properties.

Hybrid Materials in Nature

Many natural materials consist of inorganic and organic building blocks distributed on the nanoscale. In most cases the inorganic part provides mechanical strength and an overall structure to the natural objects while the organic part delivers bonding between the inorganic building blocks and/or the soft tissue. Typical examples of such materials are bone, or nacre.

Development of Hybrid Materials

The first hybrid materials were the paints made from inorganic and organic components that were used thousands of years ago. Rubber is an example of the use of inorganic materials as fillers for organic polymers. The sol–gel process developed in the 1930s was one of the major driving forces what has become the broad field of inorganic–organic hybrid materials.

Classification

Hybrid materials can be classified based on the possible interactions connecting the inorganic and organic species. *Class I*hybrid materials are those that show weak interactions between the two phases, such as van der Waals, hydrogen bonding or weak electrostatic interactions. *Class II*hybrid materials are those that show strong chemical interactions between the components such as covalent bonds.

Structural properties can also be used to distinguish between various hybrid materials. An organic moiety containing a functional group that allows the attachment to an inorganic network, e.g. a trialkoxysilane group, can act as a *network modifier* because in the final structure the inorganic network is only modified by the organic group. Phenyltrialkoxysilanes are an example for such compounds; they modify the silica network in the sol–gel process via the reaction of the trialkoxysilane group without supplying additional functional groups intended to undergo further chemical reactions to the material formed. If a reactive functional group is incorporated the system is called a *network functionalizer*. The situation is different if two or three of such anchor groups modify an organic segment; this leads to materials in which the inorganic group is afterwards an integral part of the hybrid network. The latter type of system is known as *network builder*

Blends are formed if no strong chemical interactions exist between the inorganic and organic building blocks. One example for such a material is the combination of inorganic clusters or particles with organic polymers lacking a strong (e.g. covalent) interaction between the components. In this case a material is formed that consists for example of an organic polymer with entrapped discrete inorganic moieties in which, depending on the functionalities of the components, for example weak crosslinking occurs by the entrapped inorganic units through physical interactions or the inorganic components are entrapped in a crosslinked polymer matrix. If an inorganic and an organic network interpenetrate each other without strong chemical interactions, so called interpenetrating networks (IPNs) are formed, which is for example the case if a sol–gel material is formed in presence of an organic polymer or vice versa. Both materials described belong to class Ihybrids. Class IIhybrids are formed when the discrete inorganic building blocks, e.g. clusters, are covalently bonded to the organic polymers or inorganic and organic polymers are covalently connected with each other.

Distinction between Nanocomposites and Hybrid Materials

The term nanocomposite is used if one of the structural units, either the organic or the inorganic, is in a defined size range of 1–100 nm. Large molecular building blocks for hybrid materials, such as large inorganic clusters, may be of the nanometer length scale. The term nanocomposites is generally used if discrete structural units in the respective size regime are used and the term hybrid materials is more often used if the inorganic units are formed in situ by molecular precursors, for example applying sol–gel reactions. Examples of discrete inorganic units for nanocomposites are nanoparticles, nanorods and carbon nanotubes.

Advantages of Hybrid Materials over Traditional Composites

- Inorganic clusters or nanoparticles with specific optical, electronic or magnetic properties can be incorporated in organic polymer matrices.
- Contrary to pure solid state inorganic materials that often require a high temperature treatment for their processing, hybrid materials show a more polymer-like handling, either because of their large organic content or because of the formation of crosslinked inorganic networks from small molecular precursors just like in polymerisation reactions.

- Light scattering in homogeneous hybrid material can be avoided and therefore optical transparency of the resulting hybrid materials and nanocomposites can be achieved.

Synthesis

Two different approaches can be used for the formation of hybrid materials: Either well-defined preformed building blocks are applied that react with each other to form the final hybrid material in which the precursors still at least partially keep their original integrity or one or both structural units are formed from the precursors that are transformed into a new (network) structure. It is important that the interface between the inorganic and organic materials which has to be tailored to overcome serious problems in the preparation of hybrid materials. Different building blocks and approaches can be used for their preparation and these have to be adapted to bridge the differences of inorganic and organic materials.

Building Block Approach

Building blocks at least partially keep their molecular integrity throughout the material formation, which means that structural units that are present in these sources for materials formation can also be found in the final material. At the same time typical properties of these building blocks usually survive the matrix formation, which is not the case if material precursors are transferred into novel materials. Representative examples of such well-defined building blocks are modified inorganic clusters or nanoparticles with attached reactive organic groups.

Cluster compounds often consist of at least one functional group that allows an interaction with an organic matrix, for example by copolymerisation. Depending on the number of groups that can interact, these building blocks are able to modify an organic matrix (one functional group) or form partially or fully crosslinked materials (more than one group). For instance, two reactive groups can lead to the formation of chain structures. If the building blocks contain at least three reactive groups they can be used without additional molecules for the formation of a crosslinked material.

Beside the molecular building blocks mentioned, nanosized building blocks, such as particles or nanorods, can also be used to form nanocomposites. The building block approach has one large advantage compared with the in situ formation of the inorganic or organic entities: because at least one structural unit (the building block) is well-defined and usually does not undergo significant structural changes during the matrix formation, better structure–property predictions are possible. Furthermore, the building blocks can be designed in such a way to give the best performance in the materials' formation, for example good solubility of inorganic compounds in organic monomers by surface groups showing a similar polarity as the monomers.

In recent years many building blocks have been synthesized and used for the preparation of hybrid materials. Chemists can design these compounds on a molecular scale with highly sophisticated methods and the resulting systems are used for the formation of functional hybrid materials. Many future applications, in particular in nanotechnology, focus on a bottom-up approach in which complex structures are hierarchically formed by these small building blocks. This idea is also one of the driving forces of the building block approach in hybrid materials.

In Situ Formation of the Components

The in situ formation of the hybrid materials is based on the chemical transformation of the precursors used throughout materials' preparation. Typically this is the case if organic polymers are formed but also if the sol–gel process is applied to produce the inorganic component. In these cases well-defined discrete molecules are transformed to multidimensional structures, which often show totally different properties from the original precursors. Generally simple, commercially available molecules are applied and the internal structure of the final material is determined by the composition of these precursors but also by the reaction conditions. Therefore control over the latter is a crucial step in this process. Changing one parameter can often lead to two very different materials. If, for example, the inorganic species is a silica derivative formed by the sol–gel process, the change from base to acid catalysis makes a large difference because base catalysis leads to a more particle-like microstructure while acid catalysis leads to a polymer-like microstructure. Hence, the final performance of the derived materials is strongly dependent on their processing and its optimisation.

In Situ Formation of Inorganic Materials

Many of the classical inorganic solid state materials are formed using solid precursors and high temperature processes, which are often not ompatible with the presence of organic groups because they are decomposed at elevated temperatures. Hence, these high temperature processes are not suitable for the in situ formation of hybrid materials. Reactions that are employed should have more the character of classical covalent bond formation in solutions. One of the most prominent processes which fulfill these demands is the sol–gel process. However, such rather low temperature processes often do not lead to the thermodynamically most stable structure but to kinetic products, which has some implications for the structures obtained. For example low temperature derived inorganic materials are often amorphous or crystallinity is only observed on a very small length scale, i.e. the nanometer range. An example of the latter is the formation of metal nanoparticles in organic or inorganic matrices by reduction of metal salts or organometallic precursors.

Some methods of in situ formation of inorganic materials are:

- Sol-gel process
- Nonhydrolytic sol–gel process
- Sol–gel reactions of non-silicates.

Formation of Organic Polymers in Presence of Preformed Inorganic Materials

If the organic polymerisation occurs in the presence of an inorganic material to form the hybrid material one has to distinguish between several possibilities to overcome the incompatibilty of the two species. The inorganic material can either have no surface functionalisation but the bare material surface; it can be modified with nonreactive organic groups (e.g. alkyl chains); or it can contain reactive surface groups such as polymerisable functionalities. Depending on these prerequisites the material can be pretreated, for example a pure inorganic surface can be treated with surfactants or silane coupling agents to make it compatible with the organic monomers, or functional monomers can be added that react with the surface of the inorganic material. If the inorganic component has nonreactive organic groups attached to its surface and it can be dissolved in a monomer which is subsequently polymerised, the resulting material after the organic polymerisation, is a blend. In this case the inorganic component interact only weakly or not at all with the organic polymer; hence, a class Imaterial is formed.

Homogeneous materials are only obtained in this case if agglomeration of the inorganic components in the organic environment is prevented. This can be achieved if the interactions between the inorganic components and the monomers are better or at least the same as between the inorganic components. However, if no strong chemical interactions are formed, the long-term stability of a once homogeneous material is questionable because of diffusion effects in the resulting hybrid material. The stronger the respective interaction between the components, the more stable is the final material. The strongest interaction is achieved if class IImaterials are formed, for example with covalent interactions.

Hybrid Materials by Simultaneous Formation of Both Components

Simultaneous formation of the inorganic and organic polymers can result in the most homogeneous type of interpenetrating networks. Usually the precursors for the sol–gel process are mixed with monomers for the organic polymerisation and both processes are carried out at the same time with or without solvent. Applying this method, three processes are competing with each other:

(a) the kinetics of the hydrolysis and condensation forming the inorganic phase,

(b) the kinetics of the polymerisation of the organic phase, and

(c) the thermodynamics of the phase separation between the two phases.

By tailoring the kinetics of the two polymerisations in such a way that they occur simultaneously and rapidly enough, phase separation is avoided or minimised. Additional parameters such as attractive interactions between the two moieties, as described above can also be used to avoid phase separation.

One problem that also arises from the simultaneous formation of both networks is the sensitivity of many organic polymerisation processes for sol–gel conditions or the composition of the materials formed. Ionic polymerisations, for example, often interact with the precursors or intermediates formed in the sol–gel process. Therefore, they are not usually applied in these reactions.

Applications

- Decorative coatings obtained by the embedding of organic dyes in hybrid coatings.
- Scratch-resistant coatings with hydrophobic or anti-fogging properties.
- Nanocomposite based devices for electronic and optoelectronic applications including light-emitting diodes, photodiodes, solar cells, gas sensors and field effect transistors.
- Fire retardant materials for construction industry.
- Nanocomposite based dental filling materials.
- Composite electrolyte materials for applications such as solid-state lithium batteries or supercapacitors.
- Proton conducting membranes used in fuel cells.
- Antistatic / antireflection coatings
- Corrosion protection
- Porous hybrid materials.

Center of Excellence in Nanotechnology at AIT

The Center of Excellence (CoE) in Nanotechnology is located inside the Asian Institute of Technology campus. It is one among the 8 Centres of Excellence in Thailand.

The CoEN at AIT is about applied research and graduate education in the newly emerging field of Nanotechnology. The unifying concept in the center's research activities is to make use of inexpensive wet-chemical methods to fabricate innovative materials and futuristic device components. Current research activities at the CoEN focuses on dye-sensitized solar cells, piezotronic devices, gas sensors, bio-diagnostic tools, specific microbial sensors, heavy metal ion sensors for waste water, environmental mitigation through visible light photocatalysis, self-organisation of nanoparticles, and layer-by layer growth from colloidal particles, among others. Master's degree program in Nanotechnology has been launched in 2009. The center has over 30 members from 10 different countries who are carrying outcutting edge cross-disciplinary research in Nanotechnology.

Notable international collaborations are with: State University of New York, Buffalo, USA; S N Bose National Centre for Basic Science, India; Center of Photoelectrochemical Energy, Korea University, Jochiwon, South Korea; Center for Nanobioscience, Agharkar Institute, India; Inorganic Materials Laboratory, Royal Institute of Technology, Stockholm, Sweden;Swiss Federal Institute of Technology, Switzerland; Uppsala University, Sweden;University of Quebec, Canada; Leibniz Institute of New Materials, Germany; University of California, Berkley, USA; and Griffith University, Australia.

History

The Center of Excellence in Nanotechnology was started in 2006 with an aim to address the creation of knowledge in areas relevant to Thailand, its industries and its people. Activities include joint research with other local and international universities and institutes, education and training personnel in the field of nanotechnology, technology transfer and promotion of public and industrial awareness of nanotechnology. This center provides international platform for academicians and researchers from Thailand, AIT and our partnered universities worldwide to work together in partnership with the industries.

In Thailand, there are currently 8 Centres of Excellence under Thailand National Nanotechnology Center: NANOTEC: CoE Kasetsert University, Bangkok; CoE Chulalongkorn University, Bangkok; CoE King Mongkut Institute of Technology Ladkrabang, Bangkok; CoEMahidol University, Bangkok; CoE Asian Institute of Technology, Pathum Thani; CoE Prince of Songkhla University, Phuket; CoE Khon Kaen University, Khon Kaen; and CoE Chiang MaiUniversity, Chiang Mai.

The Center of Excellence in Nanotechnology at AIT was set up, jointly with the National Nanotechnology Center, Thailand of the NSTDA, at the Asian Institute of Technology campus, Thailand. Under the agreement, AIT and NANOTEC supports the center jointly. A few faculties at AIT joined hands to propose for this center to be established in the institute. The faculties came from all the three schools and the founding members were: Prof. Joydeep Dutta, Prof. W. Kanok-Nukulchai, Dr. Oleg Shipin, Late Dr. N. Coowanitwong, Dr. M. Parnichkun, Dr. Mousa M. Nazhad, Dr. S. Venkatesh and Dr. P. Herabat. A few rooms in the Chalarm Prakiat Building in Asian Institute of Technology were allocated for its operation. Initially, due to the lack of sufficient funds the center had a hard time to start performing. Prof. Joydeep Dutta, CoEN Director, recalls, "Iused to get glasses and mugs from my house to fulfill the students requirements for glasswares in the lab." Over the years the center got crowded with new equipments coming in and more researchers joining the nano-team. Eventually in September 2009, the center was officially shifted to the Outreach building. On 8 September 2009, a state-of-the-art facility of the Center of Excellence in Nanotechnology at AIT was jointly inaugurated by Prof. Said Irandoust, AIT President, and Dr. Paritud Bhandhubanyong, Director, National Metal and Materials Technology Center. The lab had bigger and larger space and was more free for the visitors from outside.

Administration

The Asian Institute of Technology governs the administrative activities of the Center of Excellence in Nanotechnology. Prof. Joydeep Dutta,is the director of CoEN.

Campus

The facilities provided for the students in AIT is freely available to the researchers and students of the center. With a clean and green campus and sporting facilities such as football grounds, tennis fields, badminton, volleyball, basketball courts, a swimming pool, a golf course etc., the students can find time for recreation and physical workouts within the campus. A two-storied library is accessible to all the members of AIT.

Academics

Doctoral research in Nanotechnology was taking place since 2003. But, Master's degree program in Nanotechnology at AIT was recently launched in August 2009. It has attracted considerable interest amongst students across Asia. The program is open to graduates with Bachelor of engineering (electrical, chemical, mechanical, industrial, telecommunications, computer engineering, electronics, and instrumentation) as well as Bachelor of Science (physics and chemistry).

Essential features of AIT's Nanotechnology degree program are: programmatic balance; processing nanostructures properties applications (P-N-P-A); interdisciplinarity; science (including biology) and technology; integrated classroom-lab experience; hands-on experiments and use of instrumentations; curriculum linked to applications; balance between theory and practice (industrial needs); and content addressing societal impact: public safety, ethics, and awareness.

Courses Offered

The aim of the 'integrated Master's Degree programme in Nanotechnology' is to address the knowledge-based industries of the 21st century that will require continuous development of their workforce in this new field as well as a technically updated management. The Nanotechnology course is designed in such a way that students from different disciplines can get easily acquainted with the subject matter. Prof. Dutta, Director, CoeN at AIT, and Prof. G. Louis Hornyak, visiting faculty, are teaching students from different backgrounds such asMaterial sciences, Chemistry, Physics, Agriculture etc. The Nanotechnology Master's Degree program is designed to be multidisciplinary covering areas such as:

- The macroscopic and microscopic world
- Physics/electronics
- Chemistry/chemical engg.
- Biotechnology
- Material sciences
- Microsystem Technology
- Instrumentation
- Environment.

The list of the courses offered under School or an Institute-Level course: *School of engineering and Technology* and Area/Field of Study: *Nanotechnology* in AIT website is given below:

- Self Assembly and molecular manufacturing (January)
- Impact of Nanotechnology on the Society (August)
- Intellectual Property Rights for Technology Development and Management (August)
- Solid State Physics for Nanotechnology (January)
- Fundamentals of Chemistry (August)
- Catalysis (August)
- Nano Thermodynamics (January)
- Nanomaterials and Nanotechnology (January)
- Colloids and Nanoparticles (August)
- Microelectronics Fabrication Technology (August)
- Microelectromechanical and Nanoelectromechanical systems (January)
- Characterisation Tools in Nanotechnology (January)
- Advanced Seminars in Nanotechnology (January).

Admissions

Application for admissions to Master's degree in Nanotechnology can be made online or by mail by visiting the AIT website. Application for admission takes only four steps:

- Step 1: Review Requirements and Procedures

- Step 2: Download Application Forms
- Step 3: Study Guidelines for Applicants
- Step 4: Explore Financial Aid.

Research: The Center of Excellence in Nanotechnology has acquired synthesis experience of nanoparticles by reactive precipitation techniques. Gold, silver, platinum, palladium, zinc sulphide, zinc oxide and silica nanoparticles are synthesized regularly at the center for a variety of applications. The platform technology of the center can be discussed as follows:

- *Nanoparticles:* gold, silver, platinum, paladium, silica, zinc oxide, zinc sulphide
- *Nanowires:* zinc oxide
- *Coating Techniques:* sol-gel coating, pyrosol coating, layer by layer organisation, Ink-Jet printing, spin coating.

The research efforts in AIT center are broadly classified into three groups:

- Environment: photocatalysis, heavy metal ion sensors, bacteria sensing, self-cleaning windows
- Agriculture and Food: E. colisensors, E. nose, smart pesticides, gas sensors
- Alternate Energy Harvesting: nano-bio solar cells,nano-energy generators

Primary objective of the center is to seek applications of nanoparticles. Research work has extended to nanowires and extensive work is being carried out especially in the environmental application and the energy harvesting side for the application of nanowires on zinc oxide. The competence achieved at the CoEN in AIT has been primarily in the synthesis of metal nanoparticles like gold, silver, platinum and palladium, and zinc compounds namely zinc oxide and zinc sulphide. Some additional work on the development of sol-gel type titanium dioxide has also taken place. Doping of semiconductor nanoparticles and studies on the defect engineering of metal oxides have been another key area of research at the center with an objective to apply the added value obtained from defect engineering into some applications.

Synthesis of Nanoparticles

The different methods employed to synthesize nanoparticles of gold, silver, zinc sulphide, zinc oxide, silica among other is briefly given below:

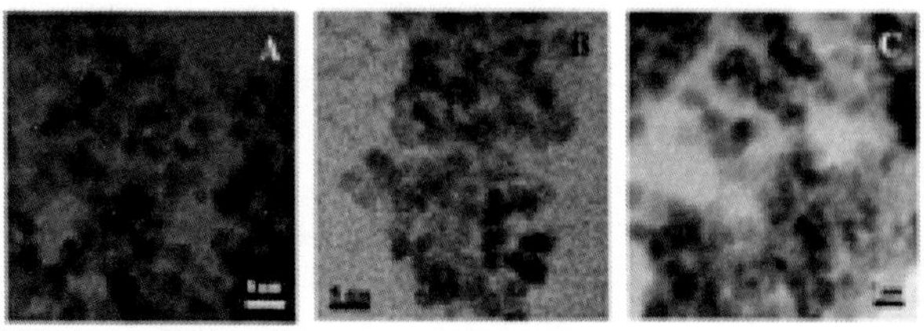

Figure: *TEM micrographs of the NPs synthesized in (A) isopropanol, (B) methonal, and (C) ethanol*

Gold Nanoparticles: Generally, gold nanoparticles in aqueous solutions are synthesized by the reduction of hydrogen tetrachloroaurate ($HAuCl_4$). To prevent the particles so formed from aggregating, often stabilising agents are added. During the process, tri-sodium citrate reduces the gold salt to metallic gold particles, which acts as seeds for continuous growth.

Silver Nanoparticles

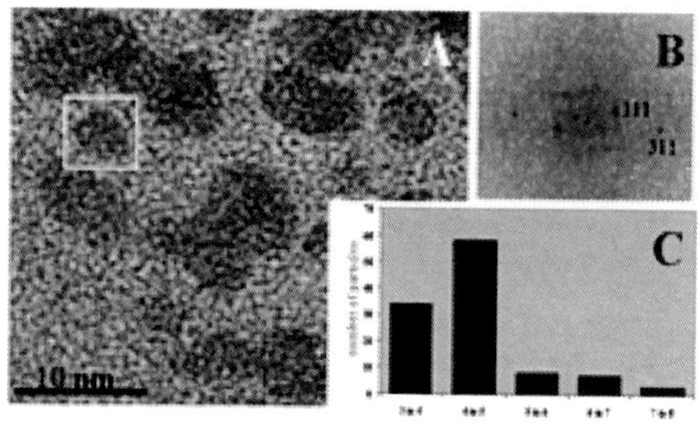

Figure: *(A) HRTEM image showing ZnS:Mn2+ NPs of size between 4 to 6 nm. Crystallites are around 2 to 3 nm. (B) Fast Fourier Transform done on the squared area (C) Particle size distribution (100 particles sampled)*

The silver nanoparticles are synthesized by chemical reduction of silver nitrate. The reducing reagent used for the synthesis of silver nanoparticles are similar to the Turkevitch method, i.e. using tri-sodium citrate. After creation of nanoparticle colloid, in order to increase the viscosity for jetting the inks, polymers such as Chitosan, Poly Vinyl Alcohol (PVA) and Poly Ethylene Glycol (PEG) are added to stabilise the solution.

ZnS Nanoparticles

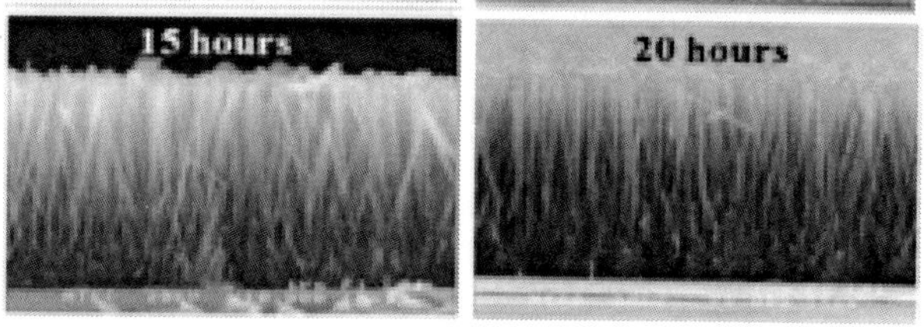

Figure: *CS view of ZnO NRs grown by hydrothermal processes for different times*

The organometallic synthesis of ZnS:Mn^{2+} nanoparticles developed is similar to the method described by Bhargava et al.

ZnO Nanoparticles

ZnO nanoparticles are synthesized in three different solvents and their optical absorptions studied to determine in which solvent the nanoparticles absorb maximum visible light in the optical band ranging between 400 nm and 700 nm. The nanoparticles synthesized in different solvents (isopropanol, methanol and ethanol) by the standard methods as explained in the experimental section yielded nanoparticles which are almost comparable in size (5 to 7 nm).

Silica Nanoparticles

Silica nanoparticles can be synthesized by sol-gel method - Stober et al. published process for synthesize silica particles by hydrolysis of tetraethyl orthosilicate (TEOS) in ethanol solution - which is catalysed by ammonia (NH_3). Monodispersed silica spheres can be achieved from hydrolysis and condensation of alkoxide silicon by the following methods:

Hydrolysis: to form silanol groups

$Si(OR)_4 + H_2O \rightarrow Si(OH)_4 + 4(R\text{-}OH)$

Condensation : to form siloxame bridges

$2Si(OH)_4 \rightarrow 2(Si\text{-}O\text{-}Si) + 4H_2O$

***Figure:** SEM micrographs showing the anisotropic ZnO NPs*

ZnO can be synthesized to various novel structures due to loose pack of tetrahedral structures, different ionic radiiof Zn^{2+} (74 pico m) and O^{2-} (140 pico m). This causes large open spaces inside the hexagonal ZnO structure and makes it possible to control the shape of nanostructured particles. The growth of ZnO structures depend on various conditions, for example, pH of solvent, temperature, duration of growth as well as nature of solvent.

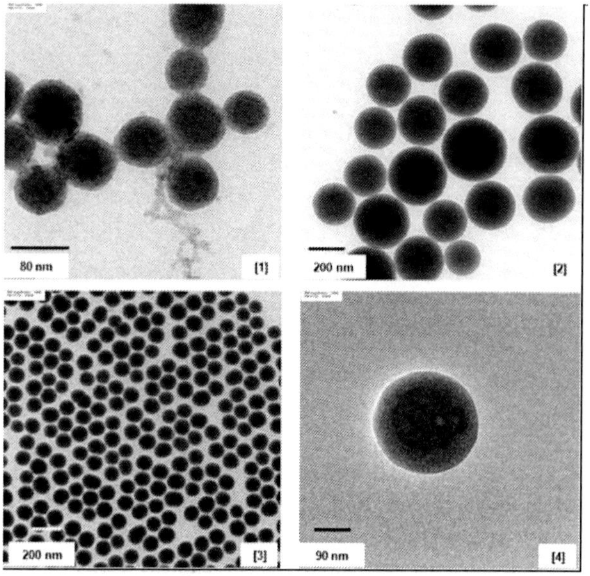

Fig. TEM images of SiO_2 NPs synthesized by Stoeber method

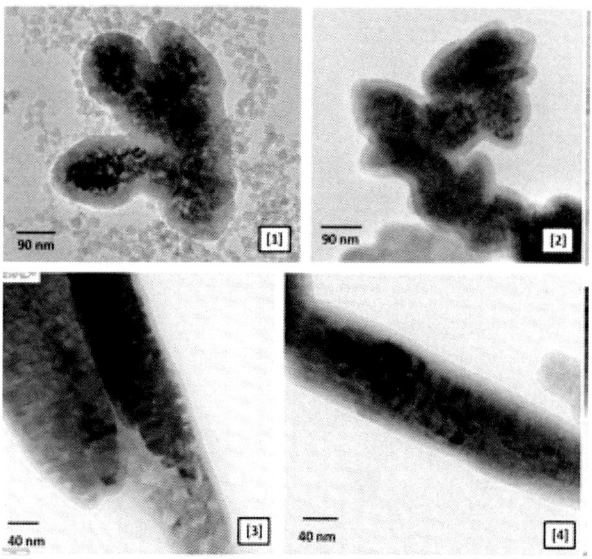

***Fig.** TEM microraphs showing the silica coated anisotropic ZnO NPs*

Besides concentration effect on particle size, it can also affect particle shape. Study from Masuda et al., synthesizing ZnO nanoparticles from zinc acetate [$Zn(CH_3COO)_2$] and ammonia precursors, shows that morphology of ZnO crystals was controlled by ratio of ammonia to zinc acetate NH_3:Zn.

Zinc acetate [$Zn(CH_3COO)_2$] solution is prepared in ethanol under vigorous stirring until zinc acetate is dissolved in ethanol in temperature around 50°C. Then ethanol is added and continuously stirred under temperature around

80°C for half an hour. After that, the solution is cooled under room temperature. NaOH in ethanol solution is prepared by vigorous stirring with 50°C, and then added to prepared zinc acetate in ethanol at room temperature. After that ZnO seeds are put in hexamethylenetetramine (HMT) solution, commonly called hexamine, for growth process. Hexamine is water soluble chemical and the growth process can be done in the temperature range 55-95°C. After ZnO nanopatilces are grown to desired shape, then the particles will be coated with silica particles by using Stoeber method. The process is carried out in ammonia-catalysed reaction of TEOS in ethanol-water solution. Polyvinylpyrrolidone (PVP) can be added optionally for stabilisation purpose. Hydrolysis reaction is initiated to form silica nanoparticles coated on ZnO nanoparticles.

ZnO Nanorods

The center is studying hydrothermal growth process of ZnO nanorods for the last few years. A typical process was carried out in a sealed chemical bath containing an equimolar solution of zinc nitrate hexahydrate and hexamethylene tetramine at a temperature of 95°C for a period of up to 20 hours. The thickness and length of the nanowires can be controlled by using different concentrations of the starting reactants and growth durations. A 0.5 mM chemical bath yielded nanowires with an average diameter of around 50 nm while a 25 mM bath resulted in wires with a thickness of up to about 1im. The length of the wires depends both on the concentration of the precursor solution as well as the growth duration, and in 20 hours, nanowires as long as 10 μm can be grown.

This center has considerable experience in the fabrication of ZnO nanorods, nanowires, and nanotubes. ZnO nanoplates that are now being used for specific applications. We have acquired considerable expertise over the last years on the control of the aspect ratio and spacing of ZnO nanorods. Recently ink-jet printing was used to make arrays of ZnO nanorods.

Applications of Nanoparticles

Antimicrobial Property of an Indoor Paint Containing Silver Nanoparticles on Escherichia Coli

The antimicrobial property of silver colloid and water-based paint embedding silver nanoparticles (AgNPs) on *Escherichia coli*(*E. coli*) were studied. Silver colloid was synthesized by method proposed herewith showed the effective inhibited growth of E. coliwith the concentration as low as 1ppm (parts per million) could completely stop the growth at E. coliconcentration above 27ppm. Further the colloids were blended with the commercial water-based paint leading to formulation of an antimicrobial paint. After adding AgNPs the paints were stable for more than a month and no change in color was observed.

The paint embedded with AgNPs showed the bactericidal effects, which is similar to silver colloids. The process to add AgNPs into the paint was simple and it was possible to integrate the current paint production lines.

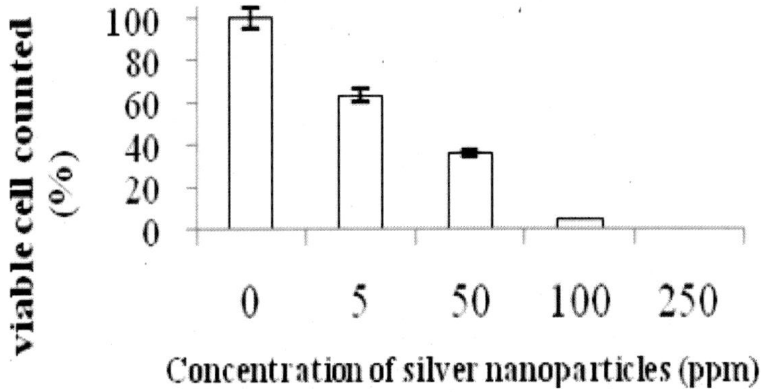

Figure: *Reduction of viable cell counted percentage of paint embedding AgNPs*

Because paint is a mixture of organic compounds, which can alter size and shape of AgNPs, some nanoparticle colloids cannot be compatibly mixed with paints. The important parameter for dispersing nanoparticles in the paint is stability of nanoparticles colloid, i.e. direct application of the zeta potential of the charged surface of nanoparticles and the pH of paint and colloid. Although the paint's recipe was modified by adding AgNPs colloids, the antimicrobial paints were stable for more than a month at room tempersture. This is because the paint and silver colloid have comparable pH and because the adding volume was small due to the high concentration of the colloid.

The pH of paint was more than 10 and the pH of the silver colloid was about 9 to 10 with zeta potential of ~50mV (milliVolt). Then, AgNPs paints were tested on antimicrobial property in order to show that the AgNPs can perform the same or better as in the colloidal form. And it turned out, the AgNPs paint showed antimicrobial property in the same way as in the colloidal form. Moreover, the inhibition of AgNPs containing paints was higher than the silver colliod due to possibly the inhibition of paints itself. In all experiments, the thickness of paint layer was minimised in order to allow bacteria to access the air. The percentage of viable cells reduced as the concentration of AgNPs in paints increased. The number of CFU (colony forming units) reduced significantly with increase in the concentration of AgNPs in paint loading 0.5ppm and higher. The optical image of cultured plates recorded after incubation showed that the paints preserved the original color and texture.

Highly Efficient Zinc Oxide Nanorod Dye Sensitized Solar Cell

ZnO nanorods grown via hydrothermal process have been used successfully as photoelectrode materials in dye-sensitized solar cells (DSSC). Such one-dimensional nanorods offer a direct conduction pathway for electron transport in the solar cell. Cell performance was evaluated from measurements of open-current voltage (Voc), short-circuit current density (Jsc), fill factor (FF), and quantum efficiency (c) vs. annealing temperature. ZnO nanorods annealed at 350°C showed the best overall performance.

Figure: *FESEM image of cross sectional view of the ZnO NRs grown by the hydrothermal process at 90°C. Inset shows the top view of the ZnO NRs*

The open circuit voltage (Voc) of a DSSC is the difference between the Fermienergy of the semiconductor (ZnO in this case) and redox potential of the electrolyte. The position of the Fermilevel changes due to the electron accumulation in the conduction band (CB) of the ZnO.

For defective nanorods, the injected electrons from dye to CB of ZnO quickly recombine in the defect sites and hence the electron density in the CB of the ZnO becomes less for defective rods. As the defect sites are removed by annealing at different temperatures (250°C, 350°C and 450°C). But at higher annealing temperature of 450°C, due to the removal of sufficient surface defects, a decrease in Voc was observed. Fewer defects in this case lead to less dye adsorption, which resulted in less electron injection to the conduction band (CB) of ZnO and thus lowered electron accumulation.

For defective ZnO nanorods, due to electron recombination, lower Jsc were observed. The maximum Jsc was obtained at an annealing temperature of 350°C. At 450°C, there was less dye adsorption, and as a result, lower Jsc was measured. The FF of the DSSC also improved with higher annealing temperature. The maximum conversion efficiency was obtained for ZnO nanorods annealed at 350°C.

Self Assembly of Gold Nanoparticles with Chitosan/PDDA Encapsulated Silica Colloidal Nanoparticles

The templated growth of organic, inorganic and hybrid networks involving self-assembly with organogels, using tuned parameters, enables us to develop innovative nanostructures. In this research work, different strategies are employed to fabricate thin films using colloidal nanoparticles of metal and insulator materials, capped with organic polymers to provide stability and electrostatic binding forces to build heterogeneous devices.

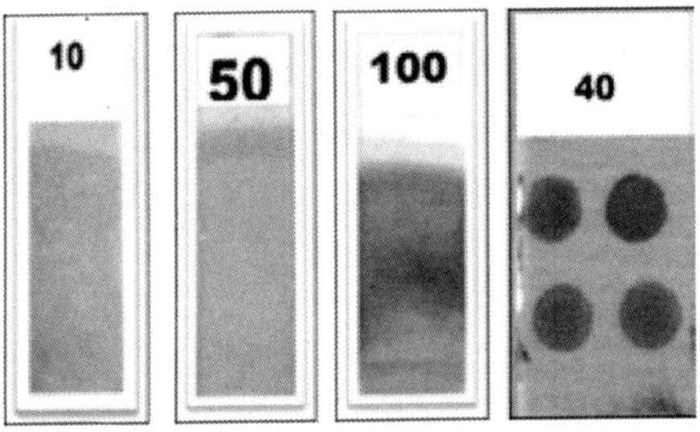

Figure: *From left to right: 10 Bi-Layers, 50 Bi-Layers, 100 Bi-Layers, and 40 Bi-Layers of CTS capped Silica & Gold*

Measurements carried out over various substrates showed repeatable optical characteristics thus demonstrating the uniformity of the deposition process. The average sizes used for the experiments of gold and silica nanoparticles were 20 nm and 100 nm, respectively. A drop of DIwater (5 iL) was placed onto the glass surface. Mean water contact angle (WCA) of the annealed glass slides were measured to be 11.9° taken through a DinoLite microscopic camera thus showing that annealing at high temperature helped to increase the hydrophilicity of the glass substrate which can improve the attachment of the first layer of electrolyte solution. Scanning Electron Microscopy (SEM) was performed. Varying the dip, dry and wash times from 5 minutes to 10 minutes resulted in an increase in the absorbance values of nanoparticles as seen in the optical absorption spectroscopy. The peak absorption values for both the Silica NPs and Gold NP s measured at wavelengths of 320 nm and 520 nm confirmed the presence as well as the uniform deposition for the both.

Defect Engineering of Zinc Oxide Nanorods for Visible Light Photocatalysis

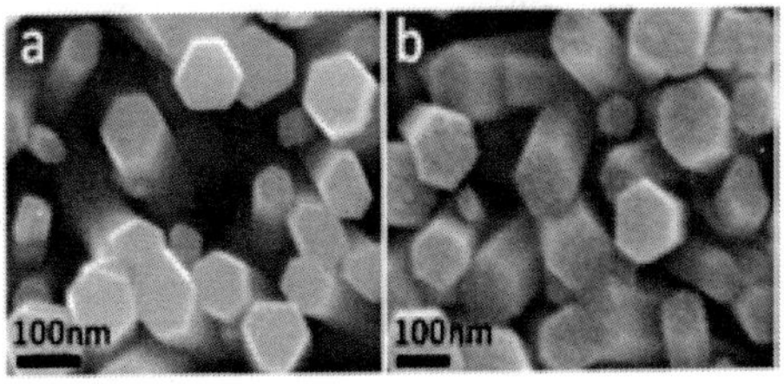

Figure: *Scanning electron micrograps of zinc oxide nanorods prepared through (a) conventional method (b) fast crystallisation*

Toxic wastes from industries are ceaselessly polluting the environment and urgent efforts are needed for the removal of harmful contaminants from soil and waste water. Wide band gapmetal oxide semiconductors like zinc oxide (ZnO) exhibit photocatalytic properties that have been exploited to degrade harmful organic contaminants into benign fragments like mineral acids. Photocatalysis is a process by which free radicals are generated in the presence of light, through the creation of electron-hole pairs that are capable of breaking up complex organic molecules. Metal oxides show enhanced photocatalytic activity with the increase in surface defects in the crystals as this can alleviate electron hole separation.

A commonly used method of incorporating defect sites or electron vacancies into the ZnO crystal is through doping using transition metals like manganese, iron, cobalt, etc. However, the center have successfully created crystal defects by controlling the crystallisation process during the growth of ZnO nanoparticles. This was possible through rapid nucleation and growth of the crystals followed by immediate quenching of the precipitation reaction. ZnO normally absorb electromagnetic waves in the ultra violet region below 370 nm (Eg= 3.37eV), but by introducing defects into its crystal lattice, a shift in the optical absorption towards the visible light band (400 nm to 700 nm) was observed. An increase in photocatalytic activity was observed due to the creation of intermediate states that inhibits photogenerated electron hole recombination, allowing redox reactions.

One dimensional nanostructures of zinc oxide like nanowires, nanorods, etc. are ideal photocatalysts because of large surface to volume ratio and inherent surface defects. Increasing the surface defects through rapid crystallisation increases the optical absorption in the visible region (e: 400-700 nm) and showed improved photocatalytic degradation of methylene blue upon illumination with white light.

Photocatalysis test was performed using a test contaminant, methylene blue [$C_{16}H_{18}N_3SC_l$] (MB) which is a heterocyclic aromatic compound, in aqueous solution. Photocatalytic degradation of (MB) results in the formation of colourless leuco methylene blue (LMB). A 10 iM solution of MB in water was put in cuvettes along with the glass slide containing ZnO nanorods. The cuvettes are placed in front of a halogen light source (500W). 72 klux of light at the sample position was measured by a luxmeter calibrated to 550 nm wavelength. Optical absorption spectra were taken after different time durations using an Ocean Optics spectrophotometer to monitor the rate of decolorisation of the test contaminant. The degradation of the dye was estimated in terms of the change in intensity at emax(~665 nm) of MTC. The degradation efficiency is calculated using the expression, where I0 is the initial absorption intensity of MTC at emax= 665 nm and Ithe intensity after photoirradiation. C0 is the initial concentration of the dye and C is the concentration after photoirradiation.

Structural defects on the polar face are clearly visible for the fast crystallised nanorods. This higher visible light absorption is indicative of the creation of mid band quasi-stable defect states to which valence band electrons can be excited with energy as low as 2.75eV. The photocatalytic degradation of MB was observed up to 60 minutes light exposure as detailed in the experimental section.

In this experiment we fixed all parameters like sample size, effective surface area available for contaminant adsorption (34.27 cm^2 for the conventional sample and 33.74 cm^2 for the fast crystallised one), intensity of irradiated light (72klux) and duration of photonic exposure (60 minutes) almost similar. It is interesting to observe that the microwave synthesized nanorods showed better photocatalytic activity as compared to the conventionally synthesized rods. This is attributed to the higher density of electron deficient sites when faster growths of the crystals were imbibed by using microwaves. These sites can trap photogenerated electrons and reduce recombinations thereby improving the photocatalytic activity.

News—August 2010

9 August 2010: National Science and Technology Fair 2010 (NSTF 2010):

> *Center of Excellence in Nanotechnology at AIT got actively involved in designing and conceptualizing the Nanotechnology Pavilion at National Science and Technology Fair 2010 being held until 22 August 2010 at Bangkok International Trade and Exhibition Center (BITEC), Bang-na, Bangkok.*

With this year's slogan "Towards a better society with Science and Technology," the National Science and Technology Fair 2010 was formally inaugurated by Dr. VirachaiVirameteekul, Minister of Science and Technology, Thailand, on 7 August 2010. The fair covered a total area of 42,000 sq. meters at the grounds of Bangkok International Trade and Exhibition Center (BITEC), Bang-na, Bangkok. With 7 ministries, 11 private enterprises, 5 major science societies, and 30 agencies from different universities participating from Thailand and more than 10 agencies participating from the overseas, it is the largest Science and Technology Fair held in Thailand. Around a million visitors, consisting of mainly school and university students from different parts of the country, are expected to visit the fair by its final day on 22 August 2010.

The participants were categorised into 9 exhibition and activity sections, namely Royal Pavilion, Ministry Pavilion, International Pavilion, University Pavilion, Science Associations, Private Enterprises, Thematic Exhibition, Activity for Youth Zones, and Academic Conference/Seminar/Workshop.

The Center of Excellence in Nanotechnology at AIT (CoEN at AIT), in collaboration with the National Science Museum, Thailand, helped in conceptualizing and designing the Nanotechnology Pavilion set up under the section: Thematic Exhibition. The exhibits at the pavilion, many of which were provided by CoEN at AIT, were designed to introduce the students to a whole new world of nanotechnology through simple demonstrations and hands-on activities. Children were busy with activities, such as making their own C_{60}fullerene (Bucky ball) model using magnetic legos, trying to understand surface to volume ratio increase with size minimisation using building blocks, playing with model explaining nanostructures etc. For the more mature audience a variety of information in the form of video for explaining the concept of scale and for demonstrating ZnO-based LPG gas sensors etc. were provided at the Pavilion.

With the growing popularity of Nanotechnology in Thailand and, in many ways, due to the active involvement of Prof. Joydeep Dutta, the Director of CoEN at AIT and the current Vice President - Academics Affairs of AIT, in promoting Nanotechnology-based research in Thailand, it was small wonder to see students from various schools and colleges taking keen interest in, what many believe is, the technology of the future.

Recreation Center - A Little Nano Band

After a hard weeks work, the Nano-group members enjoy a little recreation to ease of the stress and tension. With the initiatives of a visiting faculty to CoEN, Prof. Gabor L. Hornyak, Vice President, NanoThread, Inc., a small band named A little Nano Band was created. With professor himself on Bass or Piano, Bo Tay on rhythm guitar, Tanujjal Bora on Drums, Htet Kyaw on Tambourine and Ajaya Sapkota or Mayuree Jaisaion Vocals, the band plays covers of songs from 1960s classics to the present day pop music. The cultural diversity of the institution facilitates the band to select songs from different languages and play it to a vast group of listeners.

Zinc Oxide

Zinc oxide is an inorganic compound with the formula ZnO. It is a white powder that is insoluble in water. The powder is widely used as an additive into numerous materials and products including plastics, ceramics, glass, cement, rubber (e.g., car tires), lubricants, paints, ointments, adhesives, sealants, pigments, foods (source of Zn nutrient), batteries, ferrites, fire retardants, first aid tapes, etc. ZnO is present in the Earth's crust as the mineral zincite, however, most ZnO used commercially is synthetic.

In materials science, ZnO is a wide-bandgap semiconductor of the II-VIsemiconductor group (since zinc and oxygen belong to the 2nd and 6th groups of the periodic table, respectively). The native doping of the semiconductor (due to oxygen vacancies) is n-type. This semiconductor has several favourable properties: good transparency, high electron mobility, wide bandgap, strong room-temperature luminescence, etc. Those properties are already used in emerging applications for transparent electrodes in liquid crystal displays and in energy-saving or heat-protecting windows, and electronic applications of ZnO as thin-film transistors and light-emitting diodes are forthcoming as of 2009.

Chemical Properties

ZnO occurs as white powder known as zinc white or as the mineral zincite. The mineral usually contains a certain amount of manganese and other elements that confer yellow to red colour. Crystalline zinc oxide is thermochromic, changing from white to yellow when heated and in air reverting to white on cooling. This color change is caused by a very small loss of oxygen at high temperatures to form the non-stoichiometric $Zn_{1+x}O$, where at 800 °C, x = 0.00007.

Zinc oxide is an amphoteric oxide. It is nearly insoluble in water and alcohol, but it is soluble in (degraded by) most acids, such as hydrochloric acid:

$$ZnO + 2\ HCl \rightarrow ZnCl_2 + H_2O$$

Bases also degrade the solid to give soluble zincates:

$$ZnO + 2\ NaOH + H_2O \rightarrow Na_2[Zn(OH)_4]$$

ZnO reacts slowly with fatty acids in oils to produce the corresponding carboxylates, such asoleate or stearate. ZnO forms cement-like products when mixed with a strong aqueous solution of zinc chloride and these are best described as zinc hydroxy chlorides. This cement was used in dentistry.

ZnO also forms cement-like products when treated with phosphoric acid; related materials are used in dentistry. A major component of zinc phosphate cement produced by this reaction is hopeite, $Zn_3(PO_4)_2 \cdot 4H_2O$.

ZnO decomposes into zinc vapour and oxygen only at around 1975 °C, reflecting its considerable stability. Heating with carbon converts the oxide into the metal, which is more volatile than the oxide.

$$ZnO + C \rightarrow Zn + CO$$

Zinc oxide can react violently with aluminium and magnesium powders, with chlorinated rubber and linseed oil on heating causing fire and explosion hazard.

It reacts with hydrogen sulfide to give the sulfide: this reaction is used commercially in removing H_2S using ZnO powder (e.g., as deodorant).

$$ZnO + H_2S \rightarrow ZnS + H_2O$$

When ointments containing ZnO and water are melted and exposed to ultraviolet light, hydrogen peroxide is produced.

Physical Properties

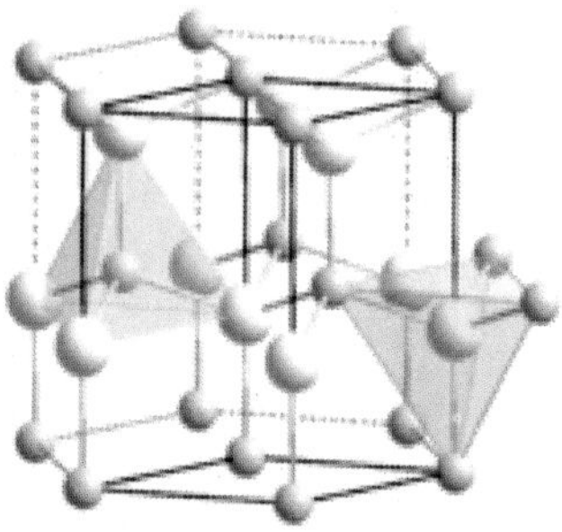

Fig. *Wurtzite structure*

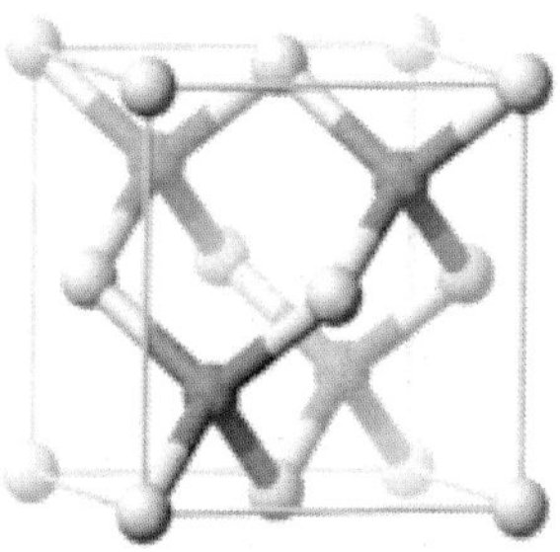

Figure: *A zincblende unit cell*

Structure

Zinc oxide crystallises in two main forms, hexagonal wurtzite and cubic zincblende. The wurtzite structure is most stable at ambient conditions and thus most common. The zincblende form can be stabilised by growing ZnO on substrates with cubic lattice structure. In both cases, the zinc and oxide centres are tetrahedral, the most characteristic geometry for Zn(II).

In addition to the wurtzite and zincblende polymorphs, ZnO can be crystallises in the rocksaltmotif at relatively high pressures about 10 GPa.

Hexagonal and zincblende polymorphs have no inversion symmetry (reflection of a crystal relative to any given point does not transform it into itself). This and other lattice symmetry properties result in piezoelectricity of the hexagonal and zincblende ZnO, and pyroelectricity of hexagonal ZnO.

The hexagonal structure has a point group 6 mm (Hermann-Mauguin notation) or C_{6v} (Schoenflies notation), and the space group is $P6_3mc$ or C_{6v}^4. The lattice constants are a = 3.25 A and c = 5.2 A; their ratio $c/a \sim 1.60$ is close to the ideal value for hexagonal cell c/a = 1.633. As in most group II-VImaterials, the bonding in ZnO is largely ionic, which explains its strong piezoelectricity. Due to the polar Zn-O bonds, zinc and oxygen planes bear electric charge (positive and negative, respectively). Therefore, to maintain electrical neutrality, those planes reconstruct at atomic level in most relative materials, but not in ZnO – its surfaces are atomically flat, stable and exhibit no reconstruction. This anomaly of ZnO is not fully explained yet.

Mechanical Properties

ZnO is a relatively soft material with approximate hardness of 4.5 on the Mohs scale. Its elastic constants are smaller than those of relevant III-V semiconductors, such as GaN. The high heat capacity and heat conductivity, low thermal expansion and high melting temperature of ZnO are beneficial for ceramics. ZnO's most stable phase being wurtzite, ZnO exhibits a very long lived optical phonon E2(low) with a lifetime as high as 133 ps at 10 K

Among the tetrahedrally bonded semiconductors, it has been stated that ZnO has the highest piezoelectric tensor, or at least one comparable to that of GaN and AlN. This property makes it a technologically important material for many piezoelectrical applications, which require a large electromechanical coupling.

Electrical Properties

ZnO has a relatively large direct band gap of ~3.3 eV at room temperature. Advantages associated with a large band gap include higher breakdown voltages, ability to sustain large electric fields, lower electronic noise, and high-temperature and high-power operation. The bandgap of ZnO can further be tuned to ~3–4 eV by its alloying with magnesium oxide or cadmium oxide.

Most ZnO has *n*-type character, even in the absence of intentional doping. Nonstoichiometry is typically the origin of n-type character, but the subject remains controversial. An alternative explanation has been proposed, based on theoretical calculations, that unintentional substitutional hydrogen impurities are responsible. Controllable n-type doping is easily achieved by substituting Zn with group-IIIelements such as Al, Ga, In or by substituting oxygen with group-VIIelements chlorine or iodine.

Reliable p-type doping of ZnO remains difficult. This problem originates from low solubility of p-type dopants and their compensation by abundant n-type impurities. This problem is observed with GaN and ZnSe. Measurement of p-type in "intrinsically" n-type material is complicated by the inhomogeneity of samples.

Current limitations to p-doping does not limit electronic and optoelectronic applications of ZnO, which usually require junctions of n-type and p-type material. Known p-type dopants include group-Ielements Li, Na, K; group-V elements N, P and As; as well as copper and silver. However, many of these form deep acceptors and do not produce significant p-type conduction at room temperature. Electron mobility of ZnO strongly varies with temperature and has a maximum of ~2000 $cm^2/(V \cdot s)$ at 80 K. Data on hole mobility are scarce with values in the range 5–30 $cm^2/(V \cdot s)$.

Production

For industrial use, ZnO is produced at levels of 10^5 tons per year by three main processes:

Indirect (French) Process

Metallic zinc is melted in a graphite crucible and vapourised at temperatures above 907 °C (typically around 1000 °C). Zinc vapor reacts with the oxygen in the air to give ZnO, accompanied by a drop in its temperature and bright luminescence. Zinc oxide particles are transported into a cooling duct and collected in a bag house. This indirect method was popularised by LeClaire (France) in 1844 and therefore is commonly known as the French process. Its product normally consists of agglomerated zinc oxide particles with an average size of 0.1 to a few micrometers. By weight, most of the world's zinc oxide is manufactured via French process.

Direct (American) Process

The direct process starts with diverse contaminated zinc composites, such as zinc ores or smelter by-products. The zinc precursors are reduced (carbothermal reduction) by heating with a source of carbon such as anthracite to produce zinc vapor, which is then oxidised as in the indirect process. Because of the lower purity of the source material, the final product is also of lower quality in the direct process as compared to the indirect one.

Wet Chemical Process

A small amount of industrial production involves wet chemical processes, which start with aqueous solutions of purified zinc salts, from which zinc carbonate or zinc hydroxide is precipitated. The precipitate is then filtered, washed, dried and calcined at temperatures around 800 °C.

Laboratory Synthesis

A large number of specialised methods exist for producing ZnO for scientific studies and niche applications. These methods can be classified by the resulting ZnO form (bulk, thin film,nanowire), temperature ("low", that is close to room temperature or "high", that is T ~ 1000 °C), process type (vapor deposition or growth from solution) and other parameters.

Large single crystals (many cubic centimeters) are usually grown by the gas transport (vapor-phase deposition), hydrothermal synthesis, or melt growth. However, because of high vapour pressure of ZnO, growth from the melt is problematic. Growth by gas transport is difficult to control, leaving the hydrothermal method as a preference. Thin films can be produced by chemical vapour deposition, metalorganic vapour phase epitaxy, electrodeposition, pulsed laser deposition, sputtering, sol-gel synthesis, atomic layer deposition, spray pyrolysis, etc.

Figure: *Synthetic ZnO crystals. Red and green color are associated with different concentrations of oxygen vacancies.*

Ordinary white powdered zinc oxide can be produced in the laboratory by electrolyzing a solution of sodium bicarbonate with a zinc anode. Zinc hydroxide and hydrogen gas are produced. The zinc hydroxide upon heating decomposes to zinc oxide.

$$Zn + 2\ H_2O \rightarrow Zn(OH)_2 + H_2$$

$$Zn(OH)_2 \rightarrow ZnO + H_2O$$

Nanophase ZnO

Nanophase ZnO can be synthesized into a variety of morphologies including nanowires, nanorods, tetrapods, nanobelts, nanoflowers, nanoparticles etc. Nanostructures can be obtained with most above-mentioned techniques, at certain conditions, and also with the vapor-liquid-solid method.

Rodlike nanostructures of ZnO can be produced via aqueous methods. They are attractive because of relatively low synthesis temperatures (<300 °C) and absence of complex vacuum setup. The synthesis is typically carried out at temperatures of about 90 °C, in an equimolar aqueous solution of zinc nitrate and hexamine, the latter providing the basic environment. Certain additives, such as polyethylene glycol or polyethylenimine, can improve the aspect ratio of the ZnO nanowires. Doping of the ZnO nanowires has been achieved by adding other metal nitrates to the growth solution. The morphology of the resulting nanostructures can be tuned by changing the parameters relating to the precursor composition (such as the zinc concentration and pH) or to the thermal treatment (such as the temperature and heating rate).

Aligned ZnO nanowires on pre-seeded silicon, glass and gallium nitride substrates have been grown in aqueous solutions using aqueous zinc salts such as Zinc nitrate and Zinc acetate in basic environments. Pre-seeding substrates with ZnO creates sites for homogeneous nucleation of ZnO crystal during the synthesis. Common pre-seeding methods include in-situ thermal decomposition of zinc acetatecrystallites, spincoating of ZnO nanoparticles and the use of various physical vapor deposition methods to deposit ZnO thin films. Pre-seeding can be performed in conjunction with top down patterning methods such as electron beam lithography and nanosphere lithography to designate nucleation sites prior to growth. Aligned ZnO nanowires can be used in dye-sensitized solar cells and field emission devices.

Applications

The applications of zinc oxide powder are numerous, and the principal ones are summarised below. Most applications exploit the reactivity of the oxide as a precursor to other zinc compounds. For material science applications, zinc oxide has high refractive index, high thermal conductivity, binding, antibacterial and UV-protection properties. Consequently, it is added into various materials and products, including plastics, ceramics, glass, cement, rubber, lubricants, paints, ointments, adhesive, sealants, pigments, foods, batteries, ferrites, fire retardants, etc.

Rubber Manufacture

About 50% of ZnO use is in the rubber industry. Zinc oxide along with stearic acid is used in the vulcanisation of rubber ZnO additive also protect rubber from fungiand UV light.

Concrete Industry

Zinc oxide is widely used for concrete manufacturing. Addition of ZnO improves the processing time and the resistance of concrete against water.

Medical

Zinc oxide as a mixture with about 0.5% iron(III) oxide (Fe_2O_3) is called calamine and is used in calamine lotion. There are also two minerals, zincite and hemimorphite, which have been historically called calamine. When mixed with eugenol, a ligand, zinc oxide eugenol is formed, which has applications as restorative and prosthodontic in dentistry.

Reflecting the basic properties of ZnO, fine particles of the oxide have deodorizing and antibacterial action and for that reason are added into various materials including cotton fabric, rubber, food packaging, etc. Enhanced antibacterial action of fine particles compared to bulk material is not intrinsic to ZnO and is observed for other materials, such as silver.

Zinc oxide is widely used to treat a variety of other skin conditions, in products such as baby powder and barrier creams to treat diaper rashes, calamine cream, anti-dandruff shampoos, and antiseptic ointments. It is also a component in tape (called "zinc oxide tape") used by athletes as a bandage to prevent soft tissue damage during workouts.

Zinc oxide can be used in ointments, creams, and lotions to protect against sunburn and other damage to the skin caused by ultraviolet light. It is the broadest spectrum UVA and UVB reflector that is approved for use as a sunscreen by the FDA, and is completely photostable. When used as an ingredient in sunscreen, zinc oxide sits on the skin's surface i.e. is not absorbed into the skin, and blocks both UVA (320–400 nm) and UVB (280–320 nm) rays of ultraviolet light. Because zinc oxide (and the other most common physical sunscreen, titanium dioxide) are not absorbed into the skin, they are nonirritating and nonallergenic.

Many sunscreens use nano-sized zinc oxide (along with nano titanium dioxide) because such small particles do not scatter light and therefore do not appear white. Although there has been concern that they might be absorbed into the skin, comprehensive reviews of the medical literature have not uncovered any risk.

Cigarette Filters

Zinc oxide is a constituent of cigarette filters for removal of selected components from tobacco smoke. A filter consisting of charcoal impregnated with zinc oxide and iron oxide removes significant amounts of HCN and H_2S from tobacco smoke without affecting its flavour.

Food Additive

Zinc oxide is added to many food products, e.g., breakfast cereals, as a source of zinc, a necessary nutrient. (Other cereals may containzinc sulfate for the same purpose.) Some prepackaged foods also include trace amounts of ZnO even if it is not intended as a nutrient.

Pigment

Zinc white is used as a pigment in paints and is more opaque than lithopone, but less opaque than titanium dioxide. It is also used in coatings for paper. Chinese white is a special grade of zinc white used in artists' pigments. It is also a main ingredient of mineral makeup.

Coatings

Paints containing zinc oxide powder have long been utilised as anticorrosive coatings for various metals. They are especially effective for galvanised Iron. The latter is difficult to protect because its reactivity with organic coatings leads to brittleness and lack of adhesion. Zinc oxide paints however, retain their flexibility and adherence on such surfaces for many years. ZnO highly n-type doped with Al, Ga or In is transparent and conductive (transparency ~90%, lowest resistivity ~10^{-4} U·cm). ZnO:Al coatings are being used for energy-saving or heat-protecting windows. The coating lets the visible part of the spectrum in but either reflects the infrared (IR) radiation back into the room (energy saving) or does not let the IR radiation into the room (heat protection), depending on which side of the window has the coating.

Various plastics, such as polyethylene naphthalate (PEN), can be protected by applying zinc oxide coating. The coating reduces the diffusion of oxygen with PEN. Zinc oxide layers can also be used on polycarbonate (PC) in outdoor applications. The coating protects PC form solar radiation and decreases the oxidation rate and photo-yellowing of PC.

Corrosion Prevention in Nuclear Reactors

Zinc oxide depleted in the zinc isotope with the atomic mass 64 is used in corrosion prevention in nuclear pressurised water reactors. The depletion is necessary, because ^{64}Zn is transformed into radioactive ^{65}Zn under irradiation by the reactor neutrons.

Potential Applications

Electronics

ZnO has wide direct band gap (3.37 eV or 375 nm at room temperature). Therefore, its most common potential applications are in laser diodes and light emitting diodes (LEDs). Some optoelectronic applications of ZnO overlap with that of GaN, which has a similar bandgap (~3.4 eV at room temperature). Compared to GaN, ZnO has a larger exciton binding energy (~60 meV, 2.4 times of the room-temperature thermal energy), which results in bright room-temperature emission from ZnO. ZnO can be combined with GaN for LED-applications. For instance as TCO layer and ZnO nanostructures provide better light outcoupling. Other properties of ZnO favourable for electronic applications include its stability to high-energy radiation and to wet chemical etching. Radiation resistance makes ZnO a suitable candidate for space applications. ZnO is currently the most promising candidate in the field of random lasers to produce an electronically pumped UV laser source.

The pointed tips of ZnO nanorods result in a strong enhancement of an electric field. Therefore, they can be used as field emitters. Aluminium-doped ZnO layers are used as a transparent electrodes. The constituents Zn and Al are much cheaper and less poisonous compared to the generally used indium tin oxide (ITO). One application which has begun to be commercially available is the use of ZnO as the front contact for solar cells or of liquid crystal displays. Transparent thin-film transistors (TTFT) can be produced with ZnO. As field-effect transistors, they even may not need a p–n junction, thus avoiding the p-type doping problem of ZnO. Some of the field-effect transistors even use ZnO nanorods as conducting channels.

Zinc Oxide Nanorod Sensor

Zinc oxide nanorod sensors are devices detecting changes in electrical current passing through zinc oxide nanowires due to adsorption of gas molecules. Selectivity to hydrogen gas was achieved by sputtering Pd clusters on the nanorod surface. The addition of Pd appears to be effective in the catalytic dissociation of hydrogen molecules into atomic hydrogen, increasing the sensitivity of the sensor device. The sensor detects hydrogen concentrations down to 10 parts per million at room temperature, whereas there is no response to oxygen.

Spintronics: ZnO has also been considered for spintronics applications: if doped with 1–10% of magnetic ions (Mn, Fe, Co, V, etc.), ZnO could become ferromagnetic, even at room temperature. Such room temperature ferromagnetism in ZnO, Mn has been observed, but it is not clear yet whether it originates from the matrix itself or from secondary oxide phases.

Piezoelectricity

The piezoelectricity in textile fibers coated in ZnO have been shown capable of fabricating "self-powered nanosystems" with everyday mechanical stress from wind or body movements. In 2008 the *Center for Nanostructure Characterisation* at the Georgia Institute of Technology reported producing an electricity generating device (called flexible charge pump generator) delivering alternating current by stretching and releasing zinc oxide nanowires. This mini-generator creates an oscillating voltage up to 45 millivolts, converting close to seven percent of the applied mechanical energy into electricity. Researchers used wires with lengths of 0.2–0.3 mm and diameters of three to five micrometers, but the device could be scaled down to smaller size.

History

It is hardly possible to trace the first usage of zinc oxide – various zinc compounds were widely used by early humans, in various processed and unprocessed forms, as a paint or medicinal ointment, but their exact composition is uncertain. The use of *pushpanjan*, probably zinc oxide, as a salve for eyes and open wounds, is mentioned in the Indian medical text the Charaka Samhita, thought to date from 500 BC or before. Zinc oxide ointment is also mentioned by the Greek physician Dioscorides (1st century AD.) Avicenna mentions zinc oxide in *The Canon of Medicine* (1025 AD), which mentioned it as a preferred treatment for a variety of skin conditions, including skin cancer. Though it is no longer used for treating skin cancer, it is still widely used to treat a variety of other skin conditions, in products such as baby powder and creams against diaper rashes, calamine cream, anti-dandruff shampoos, and antisepticointments.

The Romans produced considerable quantities of brass (an alloy of zinc and copper) as early as 200 BC by a cementation process where copper was reacted with zinc oxide. The zinc oxide is thought to have been produced by heating zinc ore in a shaft furnace. This liberated metallic zinc as a vapor, which then ascended the flue and condensed as the oxide. This process was described by Dioscorides in the 1st century AD. Zinc oxide has also been recovered from zinc mines at Zawar in India, dating from the second half of the first millennium BC. This was presumably also made in the same way and used to produce brass.

From the 12th to the 16th century zinc and zinc oxide were recognised and produced in India using a primitive form of the direct synthesis process. From India, zinc manufacture moved to China in the 17th century. In 1743, the first European zinc smelter was established in Bristol, United Kingdom. The main usage of zinc oxide (zinc white) was again paints and additive to ointments. Zinc white was accepted as a watercolour by 1834 but it did not mix well with oil. This problem was quickly solved by optimising the synthesis of ZnO. In 1845, LeClaire in Paris was producing the oil paint on a large scale, and by 1850, zinc white was being manufactured throughout Europe. The success of zinc white paint was due to its advantages over the traditional white lead: zinc white is essentially permanent in sunlight, it is not blackened by sulfur-bearing air, it is nontoxic and more economical. Because zinc white is so "clean" it is very valuable for making tints with other colours; however, it makes a rather brittle dry film when unmixed with other colours. For example, during the late 1890s and early 1900s, some artists used zinc white as a ground for their oil paintings. All those paintings developed cracks over the years. In the recent times, most zinc oxide was used in the rubber industry. In the 1970s, the second largest application of ZnO was photocopying. High-quality ZnO produced by the "French process" was added into the photocopying paper as a filler. This application was however soon displaced.

Safety

As a food additive, zinc oxide is on the U.S. FDA's generally recognised as safe, or GRAS, substances. Zinc oxide itself is nontoxic; however it is hazardous to inhale zinc oxide fumes, as generated when zinc or zinc alloys are melted and oxidised at high temperature. This problem occurs while melting brass because the melting point of brass is close to the boiling point of zinc. Exposure to zinc oxide in the air, which also occurs while welding galvanised (zinc plated) steel, can result in a nervous malady called metal fume fever. For this reason, typically galvanised steel is not welded, or the zinc is removed first.

In Popular Culture

Reflecting its many varied but relatively unobtrusive and unglamorous uses, zinc oxide was the theme of a spoof in The Kentucky Fried Movieas a short educational film *Zinc Oxide and You.*

Polyoxometalate

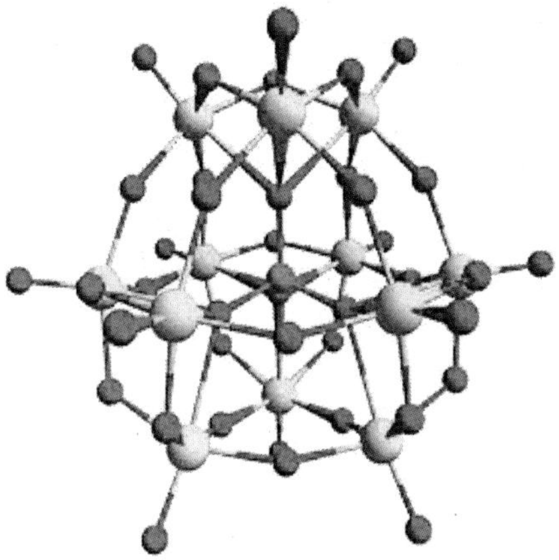

Fig. The phosphotungstate anion, an example of a polyoxometalate

In chemistry, a polyoxometalate (abbreviated POM) is a polyatomic ion, usually ananion, that consists of three or more transition metal oxyanions linked together by shared oxygen atoms to form a large, closed 3-dimensional framework. The metal atoms are usually group 5 or group 6 transition metals in their high oxdation states. In this state, their electron configuration is d^0 or d^1. Examples include vanadium(V), niobium(V), tantalum(V), molybdenum(VI), and tungsten(VI).

The framework of transition metal oxyanions may enclose one or more *hetero atoms*such as phosphorus or silicon, themselves sharing neighbouring oxygen atoms with the framework. For example, the phosphotungstate anion $[PW_{12}O_{40}]^{3"}$ consists of a framework of twelve octahedral tungsten oxyanions surrounding a central phosphategroup.

History

The first example of a polyoxometalate compound was ammonium phosphomolybdate, containing the $[PMo_{12}O_{40}]^{3"}$ anion, discovered in 1826. This anion has the same structure as the phosphotungstate anion, whose structure was determined in 1934. This structure is called the Keggin structure after its discoverer.

Following this discovery, other fundamental structures such as the Wells-Dawson ion were found, and their chemistry and applications as catalysts were determined. Recent new developments include the discovery of large, highly symmetric polyoxomolybdates such as the wheel-shaped molybdenum blueanions and spherical keplerates, numerous hybrid organic/inorganic materials that contain POM cores, new potential applications based on unusual magnetic and optical properties of some POM's, and potential medical applications such antitumour and antiviral uses.

Structure

Some structural types are found in many different compounds. The first known example of this was the Keggin ion, whose structure was found to be common to both molybdates and tungstates with different central hetero atoms. Examples of some fundamental polyoxometalate structures are shown below.

The Lindqvist ion is an *iso*-polyoxometalate, the other three are *hetero*-polyoxometalates. The Keggin and Dawson structures have tetrahedrally coordinated hetero-atoms, such as P or Si, and the Anderson structure has an octahedral central atom, such as aluminium.

Framework

The metal atoms that make up the framework, called *addenda atoms*, are typically Mo, W, and V. When more than one element is present the cluster is called a *mixed addenda* cluster.

The ligands coordinated to metal atoms that together form the bridged framework are usually oxide ions, but other elements, such as S and Br may be substituted for some of the oxide ions. A sulfur-substituted POM is called a *polyoxothiometalate*. Other ligands replacing the oxide ions have also been attested, such as nitrosyl and alkoxy groups. The typical framework building blocks are polyhedral units, with 4, 5, 6 or 7 coordinate metal centres. These units share edges and/or vertices, or, less commonly, faces (such as in the ion $CeMo_{12}O_{42}^{-8}$, which has face-shared octahedra with Mo atoms at the vertices of an icosahedron).

The most common unit for polymolybdates is the octahedral MoO_6 unit, often distorted by the Mo atom being off-centre to give one shorter Mo-O bond. Some polymolybdates contain pentagonal bipyramidal units; these are the key building blocks in the molybdenum blues.

Hetero Atoms

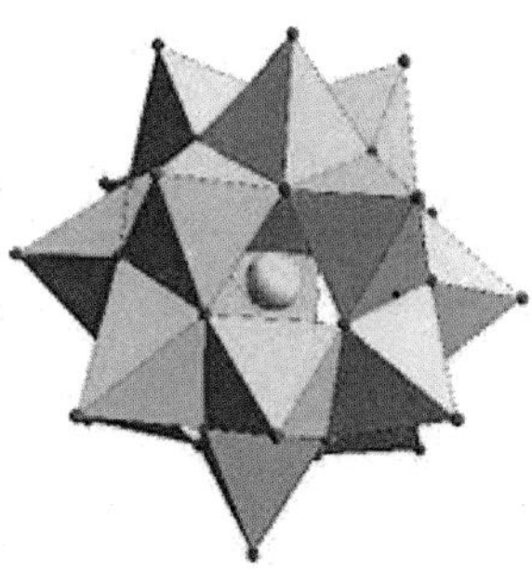

Fig. *$H_4V_{18}O_{42}$ cage containing a Cl^- ion*

Hetero atoms are present in many polyoxometalates. Many different elements can act as hetero atoms, with various coordination numbers:

- 4-coordinate (tetrahedral) in Keggin, Dawson and Lindqvist structures (e. g., PO_4, SiO_4, AsO_4)
- 6-coordinate (octahedral) in Anderson structure (e. g., $Al(OH)_6$, TeO_6)
- 8-coordinate (square antiprism) in $((CeO_8)W_{10}O_{28})^{8"}$
- 12-coordinate (icosahedral) in $(UO_{12})Mo_{12}O_{30}{}^{8"}$

The hetero atom may be located in the centre of the anion, such as in the Keggin structure, or in the center of a structural fragment, such as the two phosphorus atoms in the Dawson ion, which are central to its two symmetric fragments. Polyoxometalates bear similarities to clathrate structures. The Keggin ion, for example, can be formulated as $PO3"4 \cong M_{12}O_{36}$, and the Dawson as $(XO2"4)_2 \cong M_{18}O_{54}$. The ≅ notation denotes the physical enclosure of the left-hand side in the right-hand side. Some cage structures that contain other ions are known. For example, the vanadate cage $V_{18}O_{42}$can enclose a Cl" ion. This structure has 5-coordinate, square pyramidal vanadium units linked together.

Isomerism

Structural isomerism is common in POMs. For example, the Keggin structure has 5 isomers, which are obtained by (conceptually) rotating one or more of the four M_3O_{13} units through 60°.

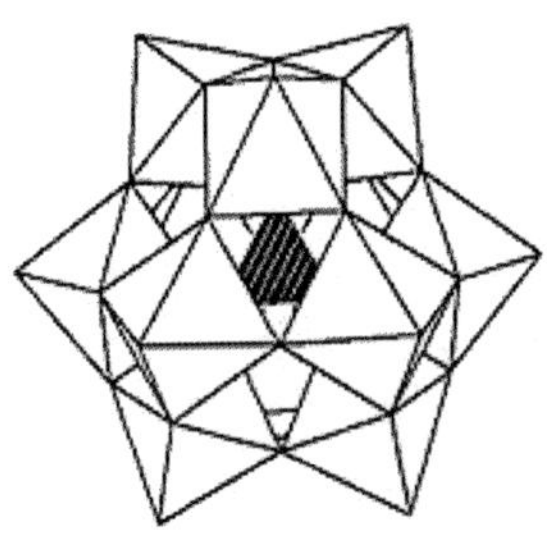

α-$XM_{12}O_{40}{}^{n-}$

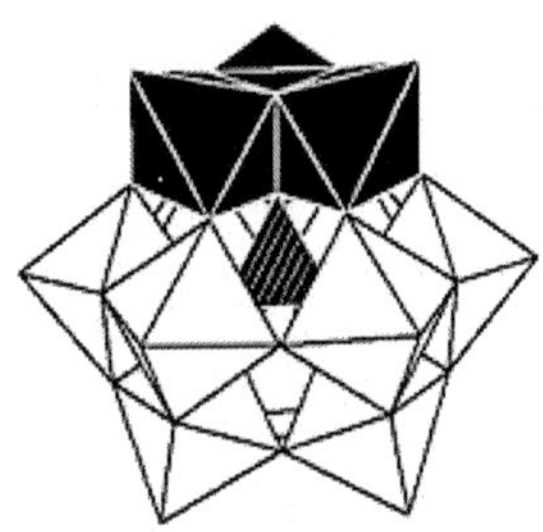

β-$XM_{12}O_{40}{}^{n-}$

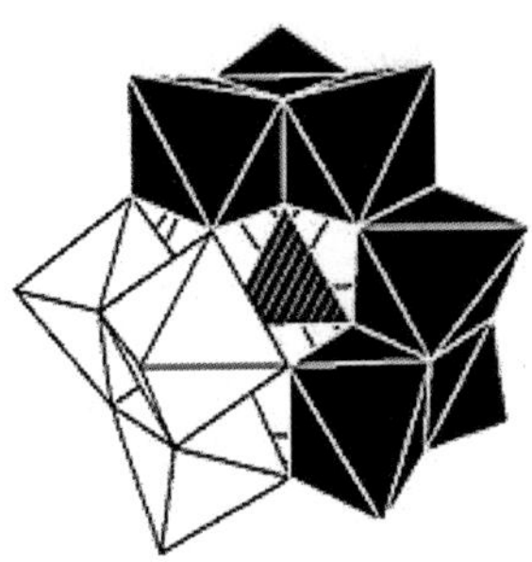

γ-$XM_{12}O_{40}{}^{n-}$

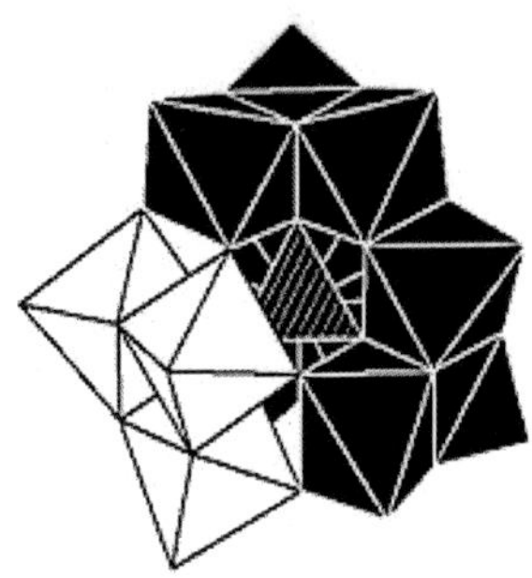

δ-$XM_{12}O_{40}{}^{n-}$

ε-$XM_{12}O_{40}^{n-}$

Fig. *The five isomers of Keggin structure.*

Lacunary Structures

The structure of some POM's are derived from a larger POM's structure by removing one or more addenda atoms and their attendant oxide ions, giving a defective structure called a *lacunary structure*. An example of a compound with a Dawson lacunary structure is $As_2W_{15}O_{56}$.

Polyoxometalates Outside Group 5 and 6 Metals

Polyoxometalates with addenda atoms outside Group 5 and 6 transition metals are known. Examples include the dodecatitanates $Ti_{12}O_{16}(OPri)_{16}$ (where OPristands for an alkoxy group) and the iron oxoalkoxometalates. These structures are also categorised as POM's, and are known as *polyoxoalkoxometalates* due to the presence of the alkoxy groups.

Properties and Applications

The huge range of size, structure and elemental composition of known polyoxometalates leads to a wide range of different properties.

The Keggin ions are well-known to be thermally stable and reversibly reduced by accepting electrons. This makes them useful as catalysts for a range of organic reactions.

Some POM's exhibit luminescence.

There have been reports on the role of weak- or non-bonding interactions on the crystal engineering of hybrid polyoxometalates.

Spherical nanoporous polyoxomolybdate-based capsules of different types containing more than 100 metal atoms reported by Achim Mullerand his group have versatile unique properties regarding their assembly into vesicles and the chemistry which can be done inside the pores and cavities. A discrete polyoxometalate Lindqvist ion of the form $W_6O_{19}^{-2}$ was successfully imaged recently for the first time within the capillary of a carbon nanotube following steric locking of the anion with the tubule. *In situ* relaxation of the anion in its equatorial plane was demonstrated.

Some structures containing transition metal atoms with unpaired electrons have unusual magnetic properties and are being investigated as possible nanocomputer storage devices.

Some potential "green" applications have been reported, such as a non-chlorine based, wood pulp bleaching process and a method of decontaminating water.

Many potential medicinal applications have been reported, such as anti-tumoural and anti-viral applications.

Carbide

Fig. *Calcium carbide.*

In chemistry, a carbide is a compound composed of carbon and a less electronegative element. Carbides can be generally classified by chemical bonding type as follows: (i) salt-like, (ii) covalent compounds, (iii) interstitial compounds, and (iv) "intermediate" transition metal carbides. Examples include calcium carbide, silicon carbide, tungsten carbide (often called simply *carbide*), and cementite, each used in key industrial applications.

Salt-like Carbides

Salt-like carbides are composed of highly electropositive elements such as the alkalimetals, alkaline earths, and group 3 metals including scandium, yttrium and lanthanum. Aluminium from group 13 forms carbides, but gallium, indium and thallium do not. These materials feature isolated carbon centres, often described as "C^{4-}", in the methanides or methides; two atom units, "C_2^{2-}" in the acetylides; and three atom units "C_3^{4-}" in the sesquicarbides. The naming of ionic carbides is not consistent and can be quite confusing.

Methanides

Carbides of this class decompose in water producing methane. Two such examples are aluminium carbide Al_4C_3 and beryllium carbi$_{\text{de Be2C.}}$

Methanides in general chemical context refers to any compound that hydrolyzes to methane, which might include also salts with hydrogenated anions such as $CH^{3"}$, CH2"2, and CH"3. However, according to IUPAC systematic naming conventions, only the last is properly called "methanide". In theory one can describe compounds that contain the methyl group, with relatively large bond polarity between the carbon and non-hydrogen atom, as salts of this anion; however in truth most such compounds, if not all, are, in fact covalent.

Acetylides

Several carbides are assumed to be salts of the acetylide anion C_2^{2-} (also called percarbide), which has a triple bond between the two carbon atoms. Alkalimetals, alkaline earth metals, and lanthanoid metals form acetylides, e.g., sodium carbide Na_2C_2, calcium carbide CaC_2, and LaC_2. Lanthanoids also form acetylides with formula M_2C_3. Metals from group 11 too tend to form acetylides, such as copper(I) acetylide and silver acetylide. Carbides of the actinide elements, which have stoichiometry MC_2 and M_2C_3, are also described as salt-like derivatives of C_2^{2-}. The C-C triple bond length ranges from 109.2 pm in CaC_2 (similar to ethyne), to 130.3 pm in LaC_2 and 134 pm in UC_2. The bonding in LaC_2has been described in terms of $La^{III}$with the extra electron delocalised into the antibonding orbital on $C_2^{2"}$, explaining the metallic conduction.

Sesquicarbides

The polyatomic ion C_3^{4-}, sometimes called sesquicarbide, is found in Li_4C_3, Mg_2C_3. The ion is linear and is isoelectronic with CO_2. The C-C distance in Mg_2C_3 is 133.2 pm. Mg2C3 yields methylacetylene, CH_3CCH, on hydrolysis which was the first indication that it may contain C_3^{4-}.

Covalent carbides

The carbides of silicon and boron are described as "covalent carbides", although virtually all compounds of carbon exhibit some covalent character. Silicon carbide has two similar crystalline forms, which are both related to the diamond structure. Boron carbide, B_4C, on the other hand has an unusual structure which includes icosahedral boron units linked by carbon atoms. In this respect boron carbide is similar to the boron rich borides. Both silicon carbide, SiC, (carborundum) and boron carbide, B_4C are very hard materials and refractory. Both materials are important industrially. Boron also forms other covalent carbides, e.g. $B_{25}C$.

Interstitial carbides

The carbides of the group 4, 5 and 6 transition metals (with the exception of chromium) are often described as interstitial compounds. These carbides have metallic properties and are refractory. Some exhibit a range of stoichiometries, e.g. titanium carbide, TiC. Titanium carbide and tungsten carbide are important industrially and are used to coat metals in cutting tools.

The longheld view is that the carbon atoms fit into octahedral interstices in a close packed metal lattice when the metal atom radius is greater than approximately 135 pm:

- When the metal atoms are cubic close packed, (ccp), then filling all of the octahedral interstices with carbon achieves 1:1 stoichiometry with the rock salt structure, (note that in rock salt, NaCl, it is the chloride anions that are cubic close packed).
- When the metal atoms are hexagonal close packed, (hcp), as the octahedral interstices lie directly opposite each other on either side of the layer of metal atoms, filling only one of these with carbon achieves 2:1 stoichiometry with the CdI_2 structure.

The following table shows actual structures of the metals and their carbides. (N.B. the body centred cubic structure adopted by vanadium, niobium, tantalum, chromium, molybdenum and tungsten is not a close packed lattice.) The notation "h/2" refers to the M_2C type structure described above, which is only an approximate description of the actual structures. The simple view that the lattice of the pure metal "absorbs" carbon atoms can be seen to be untrue as the packing of the metal atom lattice in the carbides is different from the packing in the pure metal, although it is technically correct that the carbon atoms fit into the octahedral interstices of a close-packed metal lattice.

Intermediate Transition Metal Carbides

In these carbides, the transition metal ion is smaller than the critical 135 pm, and the structures are not interstitial but are more complex. Multiple stoichiometries are common, for example iron forms a number of carbides, Fe_3C, Fe_7C_3 and Fe_2C.

The best known is cementite, Fe_3C, which is present in steels. These carbides are more reactive than the interstitial carbides, for example the carbides of Cr, Mn, Fe, Co and Niall are hydrolysed by dilute acids and sometimes by water, to give a mixture of hydrogen and hydrocarbons. These compounds share features with both the inert interstitials and the more reactive salt-like carbides.

Molecular Carbides

Metal complexes containing C_n fragments are well known. Most common are carbon-centred clusters, such as $[Au_6C(PPh_3)_6]^{2+}$. Similar species are known for the metal carbonyls and the early metal halides. Even terminal carbides have been crystallised, e.g., $CRuCl_2(P(C_6H_{11})_3)_2$.

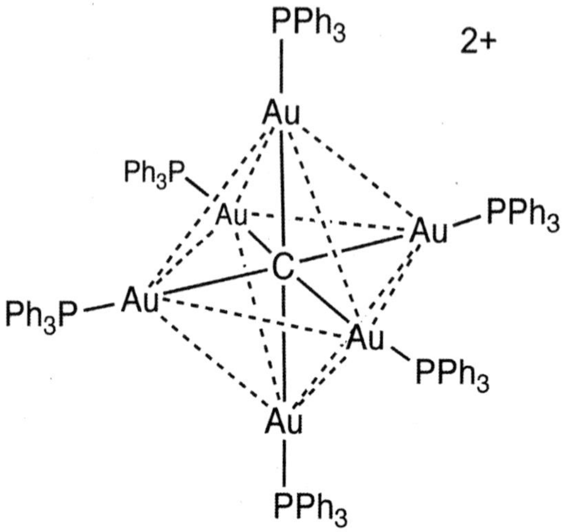

Figure: *The complex* $[Au_6C(PPh_3)_6]^{2+}$, *containing a carbon-gold core.*

Impossible Carbides

Some metals, such as lead and tin, are believed not to form carbides under any circumstances. There exists however a mixed titanium-tin carbide, which is a two-dimensional conductor. (In 2007, there were two reports of a lead carbide PbC_2, apparently of the acetylide type; but these claims have yet to be published in reviewed journals.)

Related Materials

In addition to the carbides, other groups of related carbon compounds exist:

- graphite intercalation compounds
- alkalimetal fullerides
- endohedral fullerenes, where the metal atom is encapsulated inside a fullerene molecule
- metallacarbohedrenes(met-cars) which are cluster compounds containing C_2 units.
- tunable nanoporous carbon, where gas chlorination of metallic carbides removes metal molecules to form a highly porous, near-pure carbon material capable of high-density energy storage.
- Transition metal carbene complexes.

6

Optical Characterization and Applications of Semiconductor Quantum

A semiconductor is a material which has electrical conductivity between that of a conductor such as copper and that of an insulator such as glass. Semiconductors are the foundation of modern electronics, including transistors, solar cells, light-emitting diodes (LEDs), quantum dots and digital and analog integrated circuits. The modern understanding of the properties of a semiconductor relies on quantum physics to explain the movement of electrons inside a lattice of atoms. The increasing understanding of semiconductor materials and fabrication processes has made possible continuing increases in the complexity and speed of semiconductor devices, an effect known as Moore's Law.

The conductivity of a semiconductor material increases with increasing temperature, behaviour opposite to that of a metal. Semiconductors can display a range of useful properties such as passing current more easily in one direction than the other, variable resistance, and sensitivity to light or heat. Because the conductive properties of a semiconductor material can be modified by controlled addition of impurities or by the application of electrical fields or light, devices made with semiconductors are very useful for amplification of signals, switching, and energy conversion.

Current conduction in a semiconductor occurs through the movement of free electrons and "holes", collectively known as charge carriers. Adding impurity atoms to a semiconducting material, known as "doping", greatly increases the number of charge carriers within it. When a doped semiconductor contains excess holes it is called "p-type", and when it contains excess free electrons it is known as "n-type". The semiconductor material used in devices is doped under highly controlled conditions to precisely control the location and concentration of p- and n-type dopants. A single semiconductor crystal can have multiple p- and n-type regions; the p-n junctions between these regions have many useful electronic properties.

Some of the properties of semiconductor materials were observed throughout the mid 19th and first decades of the 20th century. Development of quantum physics in turn allowed the development of the transistor in 1948. Although some pure elements and many compounds display semiconductor properties, silicon, germanium, and compounds of gallium are the most widely used in electronic devices.

Properties

Variable Conductivity: A pure semiconductor is a poor electrical conductor as a consequence of having just the right number of electrons to completely fill its valence bonds. Through various techniques (e.g., doping or gating), the semiconductor can be modified to have an excess of electrons (becoming an n-type semiconductor) or a deficiency of electrons (becoming a p-type semiconductor). In both cases, the semiconductor becomes much more conductive (the conductivity can be increased by one million fold or more). Semiconductor devices exploit this effect to shape electrical current.

Depletion

When doped semiconductors are joined to metals, to different semiconductors, and to the same semiconductor with different doping, the resulting junction often strips the electron excess or deficiency out from the semiconductor near the junction. This depletion region is rectifying (only allowing current to flow in one direction), and used to further shape electrical currents in semiconductor devices.

Energetic Electrons Travel Far

Electrons can be excited across the energy band gap of a semiconductor by various means. These electrons can carry their excess energy over distance scales of micrometres before dissipating their energy into heat – a significantly longer distance than is possible in metals. This property is essential to the operation of, e. g., bipolar junction transistors and solar cells.

Light Emission

In certain semiconductors, excited electrons can relax by emitting light instead of producing heat. These semiconductors are used in the construction of light emitting diodes and fluorescent quantum dots.

Thermal Energy Conversion

Semiconductors have large thermoelectric power factors making them useful in thermoelectric generators, as well as high thermoelectric figures of merit making them useful in thermoelectric coolers.

Materials

Figure: *Silicon crystals are the most common semiconducting materials for microelectronics.*

A large number of elements and compounds have semiconducting properties, including:

- Certain pure elements are found in Group XIV of the periodic table; the most commercially important of these elements are silicon and germanium. Silicon and germanium are used here effectively because they have 4 valence electrons in their outermost shell which gives them the ability to gain or lose electrons equally at the same time.
- Binary compounds, particularly between elements in Groups III and V, such as gallium arsenide, Groups II and VI, groups IV and VI, and between different group IV elements, e.g. silicon carbide.
- Certain ternary compounds, oxides and alloys.
- Organic semiconductors, made of organic compounds.

Most common semiconducting materials are crystalline solids, but amorphous and liquid semiconductors are also known. These include hydrogenated amorphous silicon and mixtures of arsenic, selenium and tellurium in a variety of proportions. These compounds share with better known semiconductors the properties of intermediate conductivity and a rapid variation of conductivity with temperature, as well as occasional negative resistance. Such disordered materials lack the rigid crystalline structure of conventional semiconductors such as silicon. They are generally used in thin film structures, which do not require material of higher electronic quality, being relatively insensitive to impurities and radiation damage.

Preparation of Semiconductor Materials

Semiconductors with predictable, reliable electronic properties are necessary for mass production. The level of chemical purity needed is extremely high because the presence of impurities even in very small proportions can have large effects on the properties of the material. A high degree of crystalline perfection is also required, since faults in crystal structure (such as dislocations, twins, and stacking faults) interfere with the semiconducting properties of the material. Crystalline faults are a major cause of defective semiconductor devices. The larger the crystal, the more difficult it is to achieve the necessary perfection. Current mass production processes use crystal ingots between 100 and 300 mm (4 and 10 in) in diameter which are grown as cylinders and sliced into wafers.

Because of the required level of chemical purity and the perfection of the crystal structure which are needed to make semiconductor devices, special methods have been developed to produce the initial semiconductor material. A technique for achieving high purity includes growing the crystal using the Czochralski process. An additional step that can be used to further increase purity is known as zone refining. In zone refining, part of a solid crystal is melted. The impurities tend to concentrate in the melted region, while the desired material recrystalizes leaving the solid material more pure and with fewer crystalline faults.

In manufacturing semiconductor devices involving heterojunctions between different semiconductor materials, it is often important to align the crystal lattices of the two materials by using epitaxial techniques. The lattice constant, which is the length of the repeating element of the crystal structure, is important for determining the compatibility of materials.

Biological Nanomaterials

The interface of biology and inorganic materials represents one of the fastest growing and most promising areas of nanotechnology. Since, nanomaterials can display distinct biological effects compared with bulk materials of the same chemical composition, the physico–chemical characterization of nanomaterials and the understanding of their interaction with biological media are essential for optimizing nanoparticles properties. The 'nano–bio' interface comprises dynamic physicochemical interactions, kinetics and thermodynamic exchanges between nanomaterial surfaces and the surfaces of biological components (for example small biomolecules, proteins, DNA etc.). Moreover, nanotechnology–based approaches are being explored for a variety of biomedical applications such as for drug delivery, bioimaging, tissue engineering and biosensors. A substantial number of these approaches employ nanoscale materials or bio–nanomaterials for developing unique functionalities required by the biomedical systems. For this field to evolve, we must understand the dynamic forces and molecular components that shape these interactions. In this chapter, we present our consistent efforts to explore how the interactions between nanomaterials and biological systems modify the fundamental forces that govern nanoparticle properties as well as their interactions.

Furthermore, by employing ultrafast (picosecond/femtosecond time scales) spectroscopic technique, we define the excited state dynamical pathways of the migration of electronic radiation at the bio–nanointerface.

Bionanotechnology, nanobiotechnology, and nanobiology are terms that refer to the intersection of nanotechnology and biology. Given that the subject is one that has only emerged very recently, bionanotechnology and nanobiotechnology serve as blanket terms for various related technologies.

This discipline helps to indicate the merger of biological research with various fields of nanotechnology. Concepts that are enhanced through nanobiology include: nanodevices, nanoparticles, and nanoscale phenomena that occurs within the discipline of nanotechnology. This technical approach to biology allows scientists to imagine and create systems that can be used for biological research. Biologically inspired nanotechnology uses biological systems as the inspirations for technologies not yet created. However, as with nanotechnology and biotechnology, bionanotechnology does have many potential ethical issues associated with it.

The most important objectives that are frequently found in nanobiology involve applying nanotools to relevant medical/biological problems and refining these applications. Developing new tools, such as peptoid nanosheets, for medical and biological purposes is another primary objective in nanotechnology. New nanotools are often made by refining the applications of the nanotools that are already being used. The imaging of native biomolecules, biological membranes, and tissues is also a major topic for the nanobiology researchers. Other topics concerning nanobiology include the use of cantilever array sensors and the application of nanophotonics for manipulating molecular processes in living cells.

Recently, the use of microorganisms to synthesize functional nanoparticles has been of great interest. Microorganisms can change the oxidation state of metals. These microbial processes have opened up new opportunities for us to explore novel applications, for example, the biosynthesis of metal nanomaterials. In contrast to chemical and physical methods, microbial processes for synthesizing nanomaterials can be achieved in aqueous phase under gentle and environmentally benign conditions. This approach has become an attractive focus in current green bionanotechnology research towards sustainable development.

Most of the scientific concepts in bionanotechnology are derived from other fields. Biochemical principles that are used to understand the material properties of biological systems are central in bionanotechnology because those same principles are to be used to create new technologies. Material properties and applications studied in bionanoscience include mechanical properties(e.g. deformation, adhesion, failure), electrical/electronic (e.g. electromechanical stimulation, capacitors, energy storage/batteries), optical (e.g. absorption, luminescence, photochemistry), thermal (e.g. thermomutability, thermal management), biological (e.g. how cells interact with nanomaterials, molecular flaws/defects, biosensing, biological mechanisms s.a. mechanosensing), nanoscience of disease (e.g. genetic disease, cancer, organ/tissue failure), as well as computing (e.g. DNA computing). The impact of bionanoscience, achieved through structural and mechanistic analyses of biological processes at nanoscale, is their translation into synthetic and technological applications through nanotechnology.

Nano-biotechnology takes most of its fundamentals from nanotechnology. Most of the devices designed for nano-biotechnological use are directly based on other existing nanotechnologies. Nano-biotechnology is often used to describe the overlapping multidisciplinary activities associated with biosensors, particularly where photonics, chemistry, biology, biophysics, nano-medicine, and engineering converge. Measurement in biology using wave guide techniques, such as dual polarization interferometry, are another example.

Applications

Applications of bionanotechnology are extremely widespread. Insofar as the distinction holds, nanobiotechnology is much more commonplace in that it simply provides more tools for the study of biology. Bionanotechnology, on the other hand, promises to recreate biological mechanisms and pathways in a form that is useful in other ways.

Nanobiotechnology

Nanobiotechnology (sometimes referred to as nanobiology) is best described as helping modern medicine progress from treating symptoms to generating cures and regenerating biological tissues. Three American patients have received whole cultured bladders with the help of doctors who use nanobiology techniques in their practice. Also, it has been demonstrated in animal studies that a uterus can be grown outside the body and then placed in the body in order to produce a baby. Stem cell treatments have been used to fix diseases that are found in the human heart and are in clinical trials in the United States.

There is also funding for research into allowing people to have new limbs without having to resort to prosthesis. Artificial proteins might also become available to manufacture without the need for harsh chemicals and expensive machines. It has even been surmised that by the year 2055, computers may be made out of biochemicals and organic salts.

Another example of current nanobiotechnological research involves nanospheres coated with fluorescent polymers. Researchers are seeking to design polymers whose fluorescence is quenched when they encounter specific molecules. Different polymers would detect different metabolites. The polymer-coated spheres could become part of new biological assays, and the technology might someday lead to particles which could be introduced into the human body to track down metabolites associated with tumors and other health problems. Another example, from a different perspective, would be evaluation and therapy at the nanoscopic level, i.e. the treatment of Nanobacteria (25-200 nm sized) as is done by NanoBiotech Pharma.

While nanobiology is in its infancy, there are a lot of promising methods that will rely on nanobiology in the future. Biological systems are inherently nano in scale; nanoscience must merge with biology in order to deliver biomacromolecules and molecular machines that are similar to nature. Controlling and mimicking the devices and processes that are constructed from molecules is a tremendous challenge to face the converging disciplines of nanotechnology. All living things, including humans, can be considered to be nanofoundries. Natural evolution has optimized the "natural" form of nanobiology over millions of years. In the 21st century, humans have developed the technology to artificially tap into nanobiology. This process is best described as "organic merging with synthetic." Colonies of live neurons can live together on a biochip device; according to research from Dr. Gunther Gross at the University of North Texas. Self-assembling nanotubes have the ability to be used as a structural system. They would be composed together with rhodopsins; which would facilitate the optical computing process and help with the storage of biological materials. DNA (as the software for all living things) can be used as a structural proteomic system - a logical component for molecular computing. Ned Seeman - a researcher at New York University - along with other researchers are currently researching concepts that are similar to each other.

Bionanotechnology

DNA nanotechnology is one important example of bionanotechnology. The utilisation of the inherent properties of nucleic acids like DNA to create useful materials is a promising area of modern research. Another important area of research involves taking advantage of membrane properties to generate synthetic membranes. Protein folding studies provide a third important avenue of research, but one that has been largely inhibited by our inability to predict protein folding with a sufficiently high degree of accuracy. Given the myriad uses that biological systems have for proteins, though, research into understanding protein folding is of high importance and could prove fruitful for bionanotechnology in the future. Lipid nanotechnology is another major area of research in bionanotechnology, where physico-chemical properties of lipids such as their antifouling and self-assembly is exploited to build nanodevices with applications in medicine and engineering.

Tools

This field relies on a variety of research methods, including experimental tools (e.g. imaging, characterization via AFM/optical tweezers etc.), x-ray diffraction based tools, synthesis via self-assembly, characterization of self-assembly (using e.g. dual polarization interferometry, recombinant DNA methods, etc.), theory (e.g. statistical mechanics, nanomechanics, etc.), as well as computational approaches (bottom-up multi-scale simulation, supercomputing).

Nanofabrication and Nanoelectronics

Nanofabrication is the group of techniques that allows scientists and engineers to build structures and devices at the atomic scale. Traditional top-down nanofabrication consists of carving nanoscale materials from a bulk structure through chemical means or by using beams of electrons or ions to strip away layers of material. Bottom-up methods create structures by adding atomic layers one at a time by deposition or by molecular or nanoparticle self-assembly. Penn State strengths in nanofabrication include complex ferroelectric oxide materials and devices, nanoelectronics, infrared materials and devices, MEMS, and self-assembly.

Impact on Chemistry and Related Technology

Nanochemistry or Nanotechnology related with the production and the reactions of nanoparticles and their compounds. It is concerned with the unique properties associated with assemblies of atoms or molecules on a scale between that of the individual building blocks and the bulk material (from 1 to 1000 nm), At this level, quantum effects can be significant, and also new ways of carrying out chemical reactions become possible. This science use methodologies from the synthetic chemistry and the material's chemistry to obtain nanomaterials with specific sizes, shapes, surface properties, defects, self-assembly properties, designed to accomplish specific functions and uses.

Applications

The applications of nanochemistry have a wide range which covers from the semi-conductors electronics, to medicine. Nanochemistry uses semi-conductors that only conduct electricity in specific conditions. As the semiconductors are much smaller than normal conductors the product can be much smaller. There is evidence certain nanoparticles of silver are useful to inhibit some viruses and bacteria.

Nanochemistry is being used to build high-tech armor and military weapons and for military uses. Nanochemistry could also be used for Chemical Warfare. Nanochemistry is also used to make windows that clean themselves, along with bicycles that are 1000 times stronger than steel but lighter than metal ones. The most productive piece of Nanochemistry is Carbon Nanotubes which are very dense and light when made into materials such as bicycles. Nanochemistry could be very useful in the future. The central role of chemistry in nanoscience is evident from the past 100 years of development of the chemical toolbox. Chemists today are capable of designing and synthesizing almost any imaginable compound; organic or inorganic.

Whereas chemistry has traditionally focused on basic understanding of relationship between structure and reaction, nanochemistry seeks to define and exploit the laws that govern self-assembly, quantum effects, and other fundamental nano scale phenomena. In that way chemists can make a strong and very important contribution to the broad interdisciplinary field of Nanoscience because chemists can make new systems and study how properties change when the building blocks are changed in a systematic way.

This preparative approach towards elucidation of fundamental scientific questions has proven tremendously successful in the past.

The overall aim of the nanochemistry research is to extend this approach, combined with new structural, optical and electronic tools, to reveal secrets of nanoscale assemblies. In particular we wish to develop our understanding of fundamental principles of chemical and biological self-organisation as well as the intrinsic properties of new nanosystems with intriguing chemical, biological, or physical properties.

Key areas of interest are molecular electronics, bionanoscience, and sensing.

Nanostructures in Electronics

Nanoelectronics refer to the use of nanotechnology in electronic components. The term covers a diverse set of devices and materials, with the common characteristic that they are so small that inter-atomic interactions and quantum mechanical properties need to be studied extensively. Some of these candidates include: hybrid molecular/semiconductor electronics, one-dimensional nanotubes/nanowires, or advanced molecular electronics. Recent silicon CMOS technology generations, such as the 22nm node, are already within this regime. Nanoelectronics are sometimes considered as disruptive technology because present candidates are significantly different from traditional transistors.

Nanolithography

Nanolithography is the branch of nanotechnology concerned with the study and application of fabricating nanometre-scale structures, meaning patterns with at least one lateral dimension between the size of an individual atom and approximately 100 nm. Nanolithography is used during the fabrication of leading-edge semiconductor integrated circuits (nanocircuitry) or nanoelectromechanical systems (NEMS).

As of 2007, nanolithography is a very active area of research in academia and in industry.

Nanostructured Materials for Magnetic Recording

Magnetic storage (or magnetic recording) is the storage of data on a magnetized medium. Magnetic storage uses different patterns of magnetization in a magnetizable material to store data and is a form of non-volatile memory. The information is accessed using one or more read/write heads.

As of 2013, magnetic storage media, primarily hard disks, are widely used to store computer data as well as audio and video signals. In the field of computing, the term *magnetic storage* is preferred and in the field of audio and video production, the term *magnetic recording* is more commonly used. The distinction is less technical and more a matter of preference. Other examples of magnetic storage media include floppy disks, magnetic recording tape, and magnetic stripes on credit cards.

Design

Figure: *Hard drives use magnetic memory to store giga- and terabytes of data in computers.*

Information is written to and read from the storage medium as it moves past devices called read-and-write heads that operate very close (often tens of nanometres) over the magnetic surface. The read-and-write head is used to detect and modify the magnetization of the material immediately under it. There are two magnetic polarities, each of which is used to represent either 0 or 1.

The magnetic surface is conceptually divided into many small sub-micrometre-sized magnetic regions, referred to as magnetic domains, (although these are not magnetic domains in a rigorous physical sense), each of which has a mostly uniform magnetization. Due to the polycrystalline nature of the magnetic material each of these magnetic regions is composed of a few hundred magnetic grains. Magnetic grains are typically 10 nm in size and each form a single true magnetic domain. Each magnetic region in total forms a magnetic dipole which generates a magnetic field. In older hard disk drive (HDD) designs the regions were oriented horizontally and parallel to the disk surface, but beginning about 2005, the orientation was changed to perpendicular to allow for closer magnetic domain spacing.

For reliable storage of data, the recording material needs to resist self-demagnetization, which occurs when the magnetic domains repel each other. Magnetic domains written too densely together to a weakly magnetizable material will degrade over time due to rotation of the magnetic moment one or more domains to cancel out these forces. The domains rotate sideways to a halfway position that weakens the readability of the domain and relieves the magnetic stresses. Older hard disk drives used iron(III) oxide as the magnetic material, but current disks use a cobalt-based alloy.

A write head magnetizes a region by generating a strong local magnetic field, and a read head detects the magnetization of the regions. Early HDDs used an electromagnet both to magnetize the region and to then read its magnetic field by using electromagnetic induction. Later versions of inductive heads included metal in Gap (MIG) heads and thin film heads. As data density increased, read heads using magnetoresistance (MR) came into use; the electrical resistance of the head changed according to the strength of the magnetism from the platter. Later development made use of spintronics; in read heads, the magnetoresistive effect was much greater than in earlier types, and was dubbed "giant" magnetoresistance (GMR). In today's heads, the read and write elements are separate, but in close proximity, on the head portion of an actuator arm. The read element is typically magneto-resistive while the write element is typically thin-film inductive.

The heads are kept from contacting the platter surface by the air that is extremely close to the platter; that air moves at or near the platter speed. The record and playback head are mounted on a block called a slider, and the surface next to the platter is shaped to keep it just barely out of contact. This forms a type of air bearing.

Magnetic Recording Classes

Analog Recording: Analog recording is based on the fact that remnant magnetization of a given material depends on the magnitude of the applied field. The magnetic material is normally in the form of tape, with the tape in its blank form being initially demagnetized.

When recording, the tape runs at a constant speed. The writing head magnetizes the tape with current proportional to the signal. A magnetization distribution is achieved along the magnetic tape. Finally, the distribution of the magnetization can be read out, reproducing the original signal. The magnetic tape is typically made by embedding magnetic particles in a plastic binder on polyester film tape. The commonly used magnetic particles are Iron oxide particles or Chromium oxide and metal particles with size of 0.5 micrometres. Analog recording was very popular in audio and video recording. In the past 20 years, however, tape recording has been gradually replaced by digital recording.

Digital Recording

Instead of creating a magnetization distribution in analog recording, digital recording only needs two stable magnetic states, which are the +Ms and -Ms on the hysteresis loop. Examples of digital recording are floppy disks and hard disk drives (HDDs). Digital recording has also been carried out on tapes. Recording media on HDDs use a stack of thin films to store information and a read/write head to read and write information to and from the media; various developments have been carried out in the area of used materials.

Magneto-Optical Recording

Magneto-optical recording writes/reads optically. When writing, the magnetic medium is heated locally by a laser, which induces a rapid decrease of coercive field. Then, a small magnetic field can be used to switch the magnetization. The reading process is based on magneto-optical Kerr effect. The magnetic medium are typically amorphous R-FeCo thin film (R being a rare earth element). Magneto-optical recording is not very popular. One famous example is Minidisc developed by Sony.

Domain Propagation Memory

Domain propagation memory is also called bubble memory. The basic idea is to control domain wall motion in a magnetic medium that is free of microstructure. Bubble refers to a stable cylindrical domain. Data is then recorded by the presence/absence of a bubble domain. Domain propagation memory has high insensitivity to shock and vibration, so its application is usually in space and aeronautics.

Technical Details

Access Method

Magnetic storage media can be classified as either sequential access memory or random access memory although in some cases the distinction is not perfectly clear. The access time can be defined as the average time needed to gain access to stored records. In the case of magnetic wire, the read/write head only covers a very small part of the recording surface at any given time. Accessing different parts of the wire involves winding the wire forward or backward until the point of interest is found. The time to access this point depends on how far away it is from the starting point. The case of ferrite-core memory is the opposite. Every core location is immediately accessible at any given time.

Hard disks and modern linear serpentine tape drives do not precisely fit into either category. Both have many parallel tracks across the width of the media and the read/write heads take time to switch between tracks and to scan within tracks. Different spots on the storage media take different amounts of time to access. For a hard disk this time is typically less than 10 ms, but tapes might take as much as 100 s.

Current Usage

As of 2011, common uses of magnetic storage media are for computer data mass storage on hard disks and the recording of analog audio and video works on analog tape. Since much of audio and video production is moving to digital systems, the usage of hard disks is expected to increase at the expense of analog tape. Digital tape and tape libraries are popular for the high capacity data storage of archives and backups. Floppy disks see some marginal usage, particularly in dealing with older computer systems and software. Magnetic storage is also widely used in some specific applications, such as bank cheques (MICR) and credit/debit cards (mag stripes).

Future

A new type of magnetic storage, called Magnetoresistive Random Access Memory or MRAM, is being produced that stores data in magnetic bits based on the tunnel magnetoresistance (TMR) effect. Its advantage is non-volatility, low power usage, and good shock robustness. The 1st generation that was developed was produced by Everspin Technologies, and utilised field induced writing. The 2nd generation is being developed through two approaches: Thermal Assisted Switching (TAS) which is currently being developed by Crocus Technology, and Spin Torque Transfer (STT) on which Crocus, Hynix, IBM, and several other companies are working. However, with storage density and capacity orders of magnitude smaller than an HDD, MRAM is useful in applications where moderate amounts of storage with a need for very frequent updates are required, which flash memory cannot support due to its limited write endurance.

Synthesis of Semiconductor Nanoclusters

The term clusters denotes small, multiatom particles. As a rule of thumb, any particle of somewhere between 3 and 3×10^7 atoms is considered a cluster. Two-atom particles are sometimes considered clusters as well. The term can also refer to the organisation of protons and neutrons within an atomic nucleus, e.g. the Alpha particle (also known as "α-cluster"), consisting of two protons and two neutrons (as in a helium nucleus). Although first reports of cluster species date back already to the 1940s, Cluster science emerged as a separate direction of research in the 1980s, One purpose of the research was to study the gradual development of collective phenomena which characterize a bulk solid. These are for example the colour of a body, its electrical conductivity, its ability to absorb or reflect light, and magnetic phenomena such as ferro-, ferri-, or antiferromagnetism. These are typical collective phenomena which only develop in an aggregate of a large number of atoms.

It was found that collective phenomena break down for very small cluster sizes. It turned out, for example, that small clusters of a ferromagnetic material are super-paramagnetic rather than ferromagnetic. Paramagnetism is not a collective phenomenon, which means that the ferromagnetism of the macrostate was not conserved by going into the nanostate. The question then was asked for example, "How many atoms do we need in order to obtain the collective metallic or magnetic properties of a solid?" Soon after the first cluster sources had been developed in 1980, an ever larger community of cluster scientists was involved in such studies.

This development led to the discovery of fullerenes in 1986 and carbon nanotubes a few years later.

In science, a lot is known about properties of the gas phase; however, comparatively little is known about the condensed phases (the liquid phase and solid phase.) The study of clusters attempts to bridge this gap of knowledge by clustering atoms together and studying their characteristics. If enough atoms were clustered together, eventually one would obtain a liquid or solid.

The study of atomic and molecular clusters also benefits the developing field of nanotechnology. If new materials are to be made out of nanoscale particles, such as nanocatalysts and quantum computers, the properties of the nanoscale particles (the clusters) must first be understood.

Nanoclusters constitute an intermediate state of matter between molecules and solids. Nanocluster sizes range from sub-nanometre to about 10 nanometres in diameter and are of technological interest in numerous areas of applied science (e.g. materials science, catalysis, (opto)electronics). Often one differentiates between: (i) nanoclusters consisting of up to a couple of hundred atoms, and (ii) larger aggregates containing 10^3 or more atoms which are more often called nanoparticles. The properties of nanoparticles gradually approach those of bulk materials or extended surfaces, i.e. are scalable with size. Nanoclusters, however, have properties and structures which are very sensitive to their composition and size (i.e. "every atom counts") which can lead to new and interesting properties not realised in the corresponding bulk material.

Formation of Nanostructures by Mechanical Attrition

Mechanical attrition by high energy ball milling has been developed as a versatile alternative to other processing routes including vapor evaporation, liquid quenching and chemical synthesis methods. In this process, lattice defects are produced by "pumping" energy into initially single-crystalline powder particles of typically 50 μm particle diameter. The internal refining process with a reduction of the average grain size by a factor of $10^3 - 10^4$ results from the creation and self-organisation of dislocation cell networks, small-angle grain boundaries and finally high-angle grain boundaries within the powder particles during the milling process. As a result, increases in atomic level internal strains, excess enthalpy, excess specific heat and Vickers hardness are observed far beyond the levels obtained during conventional deformation processes.

Porous Si Nanostructures

Porous silicon (p-Si) is a form of the chemical element silicon that has introduced nanoporous holes in its microstructure, rendering a large surface to volume ratio in the order of 500 m^2/cm^3.

Porous silicon was discovered by accident in 1956 by Arthur Uhlir Jr. and Ingeborg Uhlir at the Bell Labs in the U.S. At the time, the Ulhirs were in the process of developing a technique for polishing and shaping the surfaces of silicon and germanium. However, it was found that under several conditions a crude product in the form of thick black, red or brown film were formed on the surface of the material. At the time, the findings were not taken further and were only mentioned in Bell Lab's technical notes.

Despite the discovery of porous silicon in the 1950s, the scientific community was not interested in porous silicon until the late 1980s. At the time, Leigh Canham – while working at the Defence Research Agency in England – reasoned that the porous silicon may display quantum confinement effects. The intuition was followed by successful experimental results published in the 1990. In the published experiment, it was revealed that silicon wafers can emit light if subjected to electrochemical and chemical dissolution.

The published result stimulated the interest of the scientific community in its non-linear optical and electrical properties. The growing interest was evidenced in the number of published work concerning the properties and potential applications of porous silicon. In an article published in 2000, it was found that the number of published work grew exponentially in between 1991 and 1995. In 2001, a team of scientists at the Technical University of Munich inadvertently discovered that hydrogenated porous silicon reacts explosively with oxygen at cryogenic temperatures, releasing several times as much energy as an equivalent amount of TNT, at a much greater speed (an abstract of the study can be found below). Explosion occurs because the oxygen, which is in a liquid state at the necessary temperatures, is able to oxidize through the porous molecular structure of the silicon extremely rapidly, causing a very quick and efficient detonation. Although hydrogenated porous silicon would probably not be effective as a weapon, due to its functioning only at low temperatures, other uses are being explored for its explosive properties, such as providing thrust for satellites.

Fabrication of Porous Silicon

Fabrication of porous silicon may range from its initial formation through stain-etching or anodization cell, drying and storage of porous silicon and surface modification needed.

Anodization

One method of introducing pores in silicon is through the use of an anodization cell. A possible anodization cell employs platinum cathode and silicon wafer anode immersed in Hydrogen Fluoride (HF) electrolyte. Corrosion of the anode is produced by running electrical current through the cell. It is noted that the running of constant DC is usually implemented to ensure steady tip-concentration of HF resulting in a more homogeneous porosity layer although pulsed current is more appropriate for the formation of thick silicon wafers bigger than 50 μm. It was noted by Halimaoui that hydrogen evolution occurs during the formation of porous silicon. When purely aqueous HF solutions are used for the PS formation, the hydrogen bubbles stick to the surface and induce lateral and in-depth inhomogeneity The hydrogen evolution is normally treated with absolute ethanol in concentration exceeding 15%. It was found that the introduction of ethanol eliminates hydrogen and ensures complete infiltration of HF solution within the pores. Subsequently, uniform distribution of porosity and thickness is improved.

Stain Etching

It is possible to obtain porous silicon through stain-etching with hydrofluoric acid, nitric acid and water. A publication in 1957 revealed that stain films can be grown in dilute solutions of nitric acid in concentrated hydrofluoric acid.

Porous silicon formation by stain-etching is particularly attractive because of its simplicity and the presence of readily available corrosive reagents; namely nitric acid (HNO_3) and hydrogen fluoride (HF). Furthermore, stain-etching is useful if one needs to produce a very thin porous Si films. A publication in 1960 by R. J. Archer revealed that it is possible to create stain films as thin as 25 A through stain-etching with HF-HNO_3 solution.

Bottom-Up Synthesis

Porous silicon can be synthesized chemically from silicon tetrachloride according to a recent publication in Nature Communications.

Drying of Porous Silicon

Porous silicon is systematically prone to presence of cracks when the water is evaporated. The cracks are particularly evident in thick or highly porous silicon layers. The origin of the cracks has been attributed to the large capillary stress due to the minute size of the pores. In particular, it has been known that cracks will appear for porous silicon samples with thickness larger than a certain critical value. Bellet concluded that it was impossible to avoid cracking in thick porous silicon layers under normal evaporating conditions. Hence, several appropriate techniques have been developed to minimize the risk of cracks formed during drying.

Supercritical Drying

Supercritical drying is reputed to be the most efficient drying technique but is rather expensive and difficult to implement. It was first implemented by Canham in 1994 and involves superheating the liquid pore above the critical point to avoid interfacial tension.

Freeze Drying

Freeze drying procedure was first documented around 1996. After the formation of porous silicon, the sample is frozen at a temperature of about 200 K and sublimed under vacuum.

Pentane Drying

The technique uses pentane as the drying liquid instead of water. In doing so the capillary stress is reduced because pentane has a lower surface tension than water.

Slow Evaporation

Slow evaporating technique can be implemented following the water or ethanol rinsing. It was found that slow evaporation decreased the trap density

Surface Modification of Porous Silicon

The surface of porous silicon may be modified to exhibit different properties. Often, freshly etched porous silicon may be unstable due to the rate of its oxidation by the atmosphere or unsuitable for cell attachment purposes. Therefore, it can be surface modified to improve stability and cell attachment

Surface Modification Improving Stability

Following the formation of porous silicon, its surface is covered with covalently bonded hydrogen. Although the hydrogen coated surface is sufficiently stable when exposed to inert atmosphere for a short period of time, prolonged exposure render the surface prone to oxidation by atmospheric oxygen. The oxidation promotes instability in the surface and is undesirable for many applications. Thus, several methods were developed to promote the surface stability of porous silicon.

An approach that can be taken is through thermal oxidation. The process involves heating the silicon to a temperature above 1000 C to promote full oxidation of silicon. The method reportedly produced samples with good stability to aging and electronic surface passivation.

Porous silicon exhibits a high degree of biocompatibility. The large surface area enables organic molecules to adhere well. It degrades to silicic acid, which causes no harm to the body. This has opened potential applications in medicine such as a framework of the growth of bone.

Surface Modification Improving Cell Adhesion

Surface modification can also affect properties that promote cell adhesion. One particular research in 2005 studied the mammalian cell adhesion on the modified surfaces of porous silicon. The research used rat PC12 cells and Human Lens Epithelial (HLE) cells cultured for four hours on the surface modified porous silicon. Cells were then stained with vital dye FDA and observed under fluorescence microscopy. The research concluded that "amino silanisation and coating the pSi surface with collagen enhanced cell attachment and spreading".

Classification of Porous Silicon

Porosity

Porosity is defined as the fraction of void within the pSi layer and can be determined easily by weight measurement. During formation of porous silicon layer through anodization, the porosity of a wafer can be increased through increasing current density, decreasing HF concentration and thicker silicon layer. The porosity of porous silicon may range from 4% for macroporous layers to 95% for mesoporous layers. A study by Canham in 1995 found that "a 1 μm thick layer of high porosity silicon completely dissolved within a day of in-vitro exposure to a simulated body fluid". It was also found that a silicon wafer with medium to low porosity displayed more stability. Hence, the porosity of porous silicon is varied depending on its potential application areas.

Pore Size

The porosity value of silicon is a macroscopic parameter and doesn't yield any information regarding the microstructure of the layer. It is proposed that the properties of a sample are more accurately predicted if the pore size and its distribution within the sample can be obtained. Therefore, porous silicon has been divided into three categories based on the size of its pores; macroporous, mesoporous, and microporous.

Type	*Microporous*	*Mesoporous*	*Macroporous*
Pore width (Nanometer)	less than 2	Between 2 and 50	Larger than 50

Key Characteristic of Porous Silicon

Highly Controllable Properties

Porous silicon studies conducted in 1995 showed that the behaviour of porous silicon can be altered in between "bio-inert", "bioactive" and "resorbable" by varying the porosity of the silicon sample. The in-vitro study used simulated body fluid containing ion concentration similar to the human blood and tested the activities of porous silicon sample when exposed to the fluids for prolonged period of time. It was found that high porosity mesoporous layers were completely removed by the simulated body fluids within a day. In contrast, low to medium porosity microporous layers displayed more stable configurations and induced hydroxyapatite growth.

Bioactive

The first sign of porous silicon as a bioactive material was found in 1995. In the conducted study, it was found that hydroxyapatite growth was occurring on porous silicon areas. It was then suggested that "hydrated microporous Si could be a bioactive form of the semiconductor and suggest that Si itself should be seriously considered for development as a material for widespread in vivo applications." Another paper published the finding that porous silicon may be used a substrate for hydroxyapatite growth either by simple soaking process or laser-liquid-solid interaction process.

Since then, in-vitro studies have been conducted to evaluate the interaction of cells with porous silicon. A 1995 study of the interaction of B50 rat hippocampal cells with porous silicon found that B50 cells have clear preference for adhesion to porous silicon over untreated surface. The study indicated that porous silicon can be suitable for cell culturing purposes and can be used to control cell growth pattern.

Non-Toxic Waste Product

Another positive attribute of porous silicon is the degradation of porous silicon into monomeric silicic acid (SiOH4). Silicic acid is reputed to be the most natural form of element in the environment and is readily removed by kidneys. The human blood plasma contains monomeric silicic acid at levels of less than 1 mg Si/l, corresponding to the average dietary intake of 20–50 mg/day. It was proposed that the small thickness of silicon coatings presents minimal risk to a toxic concentration being reached. The proposal was supported by an experiment involving volunteers and silicic-acid drinks. It was found that concentration of the acid rose only briefly above the normal 1 mg Si/l level and was efficiently expelled by urine excretion.

Superhydrophobicity

The simple adjustment of pore morphology and geometry of porous silicon also offers a convenient way to control its wetting behaviour. Stable ultra- and superhydrophobic states on porous silicon can be fabricated and used in lab-on-a-chip, microfluidic devices for the improved surface-based bioanalysis.

Optical Properties

pSi demonstrates optical properties based on porosity and the medium inside the pores. The effective refractive index of pSi is determined by the porosity and refractive index of the medium inside the pores. If the refractive index of the medium inside pores is high, the effective refractive index of pSi will be high as well. This phenomenon causes the spectrum to shift towards longer wavelength.

Quantum Wire

In condensed matter physics, a quantum wire is an electrically conducting wire in which quantum effects influence the transport properties.

Quantum Effects

If the diameter of a wire is sufficiently small, electrons will experience quantum confinement in the transverse direction. As a result, their transverse energy will be quantized into a series of discrete values. One consequence of this quantization is that the classical formula for calculating the electrical resistance of a wire:

$$R = \rho \frac{l}{A}$$

is not valid for quantum wires (where ρ is the resistivity, l is the length, and A is the cross-sectional area of the wire).

Instead, an exact calculation of the transverse energies of the confined electrons has to be performed to calculate a wire's resistance. Following from the quantization of electron energy, the electrical conductance (the inverse of the resistance) is found to be quantized in multiples of $2e^2/h$, where e is the electron charge and h is the Planck constant. The factor of two arises from spin degeneracy. A single ballistic quantum channel (i.e. with no internal scattering) has a conductance equal to this quantum of conductance. The conductance is lower than this value in the presence of internal scattering.

The importance of the quantization is inversely proportional to the diameter of the nanowire for a given material. From material to material, it is dependent on the electronic properties, especially on the effective mass of the electrons. Physically, this means that it will depend on how conduction electrons interact with the atoms within a given material. In practice, semiconductors can show clear conductance quantization for large wire transverse dimensions (~100 nm) because the electronic modes due to confinement are spatially extended. As a result, their Fermi wavelengths are large and thus they have low energy separations. This means that they can only be resolved at cryogenic temperatures (within a few degrees of absolute zero) where the thermal energy is lower than the inter-mode energy separation.

For metals, quantization corresponding to the lowest energy states is only observed for atomic wires. Their corresponding wavelength being thus extremely small they have a very large energy separation which makes resistance quantization observable even at room temperature.

Carbon Nanotubes as Quantum Wires

The carbon nanotube is an example of a quantum wire. A metallic single-walled carbon nanotube that is sufficiently short to exhibit no internal scattering (ballistic transport) has a conductance that approaches two times the conductance quantum, $4e^2 / h$. The factor of two arises because carbon nanotubes have two spatial channels.

Quantum Point Contact

A quantum point contact (QPC) is a narrow constriction between two wide electrically conducting regions, of a width comparable to the electronic wavelength (nano- to micrometre). Quantum point contacts were first reported in 1988 by a Dutch group (Van Wees *et al.*) and, independently, by a British group (Wharam *et al.*). They are based on earlier work by the British group which showed how split gates could be used to convert a two-dimensional electron gas into one-dimension, first in silicon (Dean and Pepper) and then in gallium arsenide

Fabrication

There are several different ways of fabricating a QPC. It can be realised in a break-junction by pulling apart a piece of conductor until it breaks. The breaking point forms the point contact. In a more controlled way, quantum point contacts are formed in a two-dimensional electron gas (2DEG), e.g. in GaAs/AlGaAs heterostructures. By applying a voltage to suitably shaped gate electrodes, the electron gas can be locally depleted and many different types of conducting regions can be created in the plane of the 2DEG, among them quantum dots and quantum point contacts. Another means of creating a QPC is by positioning the tip of a scanning tunneling microscope close to the surface of a conductor.

Properties

Geometrically, a quantum point contact is a constriction in the transverse direction which presents a resistance to the motion of electrons. Applying a voltage V across the point contact induces a current to flow, the magnitude of this current is given by $I = GV$, where G is the conductance of the contact. This formula resembles Ohm's law for macroscopic resistors. However there is a fundamental difference here resulting from the small system size which requires a quantum mechanical analysis.

At low temperatures and voltages, electrons contributing to the current have a certain energy/momentum/wavelength called Fermi energy/momentum/wavelength. Much like in a waveguide, the transverse confinement in the quantum point contact results in a "quantization" of the transverse motion—the transverse motion cannot vary continuously, but has to be one of a series of discrete modes. The electron wave can only pass through the constriction if it interferes constructively, which for a given width of constriction, only happens for a certain number of modes N. The current carried by such a quantum state is the product of the velocity times the electron density. These two quantities by themselves differ from one mode to the other, but their product is mode independent. As a consequence, each state contributes the same amount e^2/h per spin direction to the total conductance

$$G = NG_Q.$$

This is a fundamental result; the conductance does not take on arbitrary values but is quantized in multiples of the conductance quantum $G_Q = 2e^2/h$, which is expressed through the electron charge e and the Planck constant h. The integer number N is determined by the width of the point contact and roughly equals the width divided by half the electron wavelength. As a function of the width of the point contact (or gate voltage in the case of GaAs/AlGaAs heterostructure devices), the conductance shows a staircase behaviour as more and more modes (or channels) contribute to the electron transport. The step-height is given by G_Q.

An external magnetic field applied to the quantum point contact lifts the spin degeneracy and leads to half-integer steps in the conductance. In addition, the number N of modes that contribute becomes smaller. For large magnetic fields, N is independent of the width of the constriction, given by the theory of the quantum Hall effect. An interesting feature, not yet fully understood, is a plateau at $0.7G_Q$, the so-called 0.7-structure.

Applications

Apart from studying fundamentals of charge transport in mesoscopic conductors, quantum point contacts can be used as extremely sensitive charge detectors. Since the conductance through the contact strongly depends on the size of the constriction, any potential fluctuation (for instance, created by other electrons) in the vicinity will influence the current through the QPC. It is possible to detect single electrons with such a scheme. In view of quantum computation in solid-state systems, QPCs may be used as readout devices for the state of a quantum bit (qubit).

Etching (Microfabrication)

Etching is used in microfabrication to chemically remove layers from the surface of a wafer during manufacturing. Etching is a critically important process module, and every wafer undergoes many etching steps before it is complete.

For many etch steps, part of the wafer is protected from the etchant by a "masking" material which resists etching. In some cases, the masking material is a photoresist which has been patterned using photolithography. Other situations require a more durable mask, such as silicon nitride.

Figures of Merit

If the etch is intended to make a cavity in a material, the depth of the cavity may be controlled approximately using the etching time and the known etch rate. More often, though, etching must entirely remove the top layer of a multilayer structure, without damaging the underlying or masking layers. The etching system's ability to do this depends on the ratio of etch rates in the two materials (*selectivity*).

Some etches undercut the masking layer and form cavities with sloping sidewalls. The distance of undercutting is called *bias*. Etchants with large bias are called *isotropic*, because they erode the substrate equally in all directions. Modern processes greatly prefer anisotropic etches, because they produce sharp, well-controlled features.

Selectivity	*Yellow: Layer to be Removed; Blue: Layer to Remain*
	1. A poorly selective etch removes the top layer, butalso attacks the underlying material.
	2. A highly selective etch leaves the underlying material unharmed.
Isotropy	Red: masking layer; yellow: layer to be removed 1. A perfectly isotropic etch produces round sidewalls.
	2. A perfectly anisotropic etch produces vertical sidewalls.

Etching Media and Technology

The two fundamental types of etchants are liquid-phase ("wet") and plasma-phase ("dry"). Each of these exists in several varieties.

Wet Etching

The first etching processes used liquid-phase ("wet") etchants. The wafer can be immersed in a bath of etchant, which must be agitated to achieve good process control. For instance, buffered hydrofluoric acid (BHF) is used commonly to etch silicon dioxide over a silicon substrate. Different specialised etchants can be used to characterise the surface etched.

Wet etchants are usually isotropic, which leads to large bias when etching thick films. They also require the disposal of large amounts of toxic waste. For these reasons, they are seldom used in state-of-the-art processes. However, the photographic developer used for photoresist resembles wet etching.

As an alternative to immersion, single wafer machines use the Bernoulli principle to employ a gas (usually, pure nitrogen) to cushion and protect one side of the wafer while etchant is applied to the other side. It can be done to either the front side or back side. The etch chemistry is dispensed on the top side when in the machine and the bottom side is not affected. This etch method is particularly effective just before "backend" processing (BEOL), where wafers are normally very much thinner after wafer backgrinding, and very sensitive to thermal or mechanical stress. Etching a thin layer of even a few micrometres will remove microcracks produced during backgrinding resulting in the wafer having dramatically increased strength and flexibility without breaking.

Anisotropic Wet Etching (Orientation Dependent Etching)

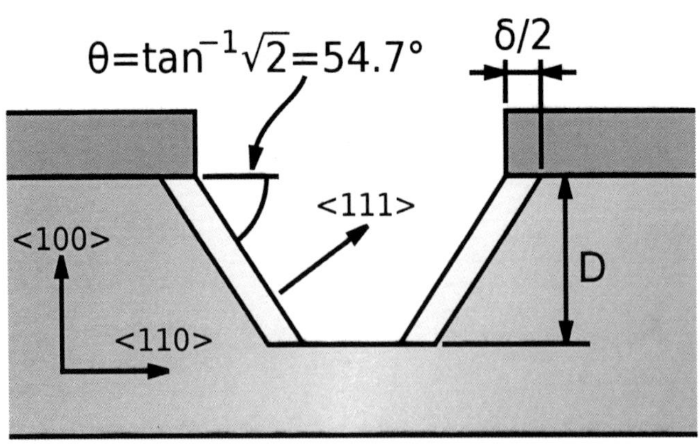

Figure: *An anisotropic wet etch on a silicon wafer creates a cavity with a trapezoidal cross-section. The bottom of the cavity is a plane, and the sides are planes. The blue material is an etch mask, and the green material is silicon.*

Some wet etchants etch crystalline materials at very different rates depending upon which crystal face is exposed. In single-crystal materials (e.g. silicon wafers), this effect can allow very high anisotropy, as shown in the figure. The term "crystallographic etching" is synonymous with "anisotropic etching along crystal planes".

However, for some non-crystal materials like glass, there are unconventional ways to etch in an anisotropic manner. The authors employs multistream laminar flow that contains etching non-etching solutions to fabricate a glass groove. The etching solution at the centre is flanked by non-etching solutions and the area contacting etching solutions is limited by the surrounding non-etching solutions. Thereby, the direction of etching is mainly vertical to the surface of glass. The SEM images demonstrate the breaking of conventional theoretical limit of aspect ratio (width/height=0.5) and contribute a two-fold improvement (width/height=1).

Several anisotropic wet etchants are available for silicon, all of them hot aqueous caustics. For instance, potassium hydroxide (KOH) displays an etch rate selectivity 400 times higher in <100> crystal directions than in <111> directions. EDP (an aqueous solution of ethylene diamine and pyrocatechol), displays a <100>/<111> selectivity of 17X, does not etch silicon dioxide as KOH does, and also displays high selectivity between lightly doped and heavily boron-doped (p-type) silicon. Use of these etchants on wafers that already contain CMOS integrated circuits requires protecting the circuitry. KOH may introduce mobile potassium ions into silicon dioxide, and EDP is highly corrosive and carcinogenic, so care is required in their use. Tetramethylammonium hydroxide (TMAH) presents a safer alternative than EDP, with a 37X selectivity between {100} and {111} planes in silicon.

Etching a (100) silicon surface through a rectangular hole in a masking material, for example a hole in a layer of silicon nitride, creates a pit with flat sloping {111}-oriented sidewalls and a flat (100)-oriented bottom. The {111}-oriented sidewalls have an angle to the surface of the wafer of:

$$\arctan\sqrt{2} = 54.7^{\circ}$$

If the etching is continued "to completion", i.e. until the flat bottom disappears, the pit becomes a trench with a V-shaped cross section. If the original rectangle was a perfect square, the pit when etched to completion displays a pyramidal shape.

The undercut, δ, under an edge of the masking material is given by:

$$\delta = \frac{\sqrt{6}D}{S} = \frac{\sqrt{6}R_{100}T}{R_{100}/R_{111}} = \sqrt{6}TR_{111},$$

where R_{xxx} is the etch rate in the <xxx> direction, T is the etch time, D is the etch depth and S is the anisotropy of the material and etchant.

Different etchants have different anisotropies. Below is a table of common anisotropic etchants for silicon:

Etchant	Operating temp (°C)	R_{100} (μm/min)	$S=R_{100}/R_{111}$	Mask materials
Ethylenediamine pyrocatechol (EDP)	110	0.47	17	SiO_2, Si_3N_4, Au, Cr, Ag, Cu
Potassium hydroxide/Isopropyl alcohol (KOH/IPA)	50	1.0	400	Si_3N_4, SiO_2 (etches at 2.8 nm/min)
Tetramethylammonium hydroxide (TMAH)	80	0.6	37	Si_3N_4, SiO_2

Plasma Etching

Modern VLSI processes avoid wet etching, and use *plasma etching* instead. Plasma etchers can operate in several modes by adjusting the parameters of the plasma. Ordinary plasma etching operates between 0.1 and 5 Torr. (This unit of pressure, commonly used in vacuum engineering, equals approximately 133.3 pascals.) The plasma produces energetic free radicals, neutrally charged, that react at the surface of the wafer. Since neutral particles attack the wafer from all angles, this process is isotropic.

Plasma etching can be isotropic, i.e., exhibiting a lateral undercut rate on a patterned surface approximately the same as its downward etch rate, or can be anisotropic, i.e., exhibiting a smaller lateral undercut rate than its downward etch rate. Such anisotropy is maximized in deep reactive ion etching. The use of the term anisotropy for plasma etching should not be conflated with the use of the same term when referring to orientation-dependent etching.

The source gas for the plasma usually contains small molecules rich in chlorine or fluorine. For instance, carbon tetrachloride (CCl_4) etches silicon and aluminium, and trifluoromethane etches silicon dioxide and silicon nitride. A plasma containing oxygen is used to oxidize ("ash") photoresist and facilitate its removal.

Ion milling, or *sputter etching*, uses lower pressures, often as low as 10^{-4} Torr (10 mPa). It bombards the wafer with energetic ions of noble gases, often Ar^+, which knock atoms from the substrate by transferring momentum. Because the etching is performed by ions, which approach the wafer approximately from one direction, this process is highly anisotropic. On the other hand, it tends to display poor selectivity. *Reactive-ion etching* (RIE) operates under conditions intermediate between sputter and plasma etching (between 10^{-3} and 10^{-1} Torr). *Deep reactive-ion etching* (DRIE) modifies the RIE technique to produce deep, narrow features.

Common Etch Processes Used in Microfabrication

Etchants for common microfabrication materials		
Material to be etched	***Wet etchants***	***Plasma etchants***
Aluminium (Al)	80% phosphoric acid (H_3PO_4) + 5% acetic acid + 5% nitric acid (HNO_3) + 10% water (H_2O) at 35–45 °C	Cl_2, CCl_4, $SiCl_4$, BCl_3
Indium tin oxide [ITO] (In_2O_3:SnO_2)	Hydrochloric acid (HCl) + nitric acid (HNO_3) + water (H_2O) (1:0.1:1) at 40 °C	
Chromium (Cr)	• "Chrome etch": ceric ammonium nitrate ($(NH_4)_2Ce(NO_3)_6$) + nitric acid (HNO_3) • Hydrochloric acid (HCl)	
Gallium Arsenide (GaAs)	• Citric Acid diluted ($C_6H_8O_7$: H_2O, 1 : 1) + Hydrogen Peroxide (H_2O_2)+ Water (H_2O)	• Cl_2, CCl_4, $SiCl_4$, BCl_3, CCl_2F_2
Gold (Au)	Aqua regia, Iodine and Potassium Iodide solution	
Molybdenum (Mo)		CF_4
Organic residues and photoresist	Piranha etch: sulfuric acid (H_2SO_4) + hydrogen peroxide (H_2O_2)	O_2 (ashing)
Platinum (Pt)	Aqua regia	
Silicon (Si)	• Nitric acid (HNO_3) + hydrofluoric acid (HF) • Potassium hydroxide (KOH) • Ethylenediamine pyrocatechol (EDP) • Tetramethylammonium hydroxide (TMAH)	• CF_4, SF_6, NF_3 • Cl_2, CCl_2F_2
Silicon dioxide (SiO_2)	• Hydrofluoric acid (HF) • Buffered oxide etch [BOE]: ammonium fluoride (NH_4F) and hydrofluoric acid (HF)	CF_4, SF_6, NF_3
Silicon nitride (Si_3N_4)	• 85% Phosphoric acid (H_3PO_4) at 180 °C (Requires SiO_2 etch mask)	CF_4, SF_6, NF_3, CHF_3
Tantalum (Ta)		CF_4
Titanium (Ti)	Hydrofluoric acid (HF)	BCl_3
Titanium nitride (TiN)	• Nitric acid (HNO_3) + hydrofluoric acid (HF) • SC1 • Buffered HF (bHF)	
Tungsten (W)	• Nitric acid (HNO_3) + hydrofluoric acid (HF) • Hydrogen Peroxide (H_2O_2)	• CF_4 • SF_6

7

Carbon and Boron Nitride Nanotubes

Carbon nanotubes (CNTs) are allotropes of carbon with a cylindrical nanostructure. Nanotubes have been constructed with length-to-diameter ratio of up to 132,000,000:1, significantly larger than for any other material. These cylindrical carbon molecules have unusual properties, which are valuable for nanotechnology, electronics, optics and other fields of materials science and technology. In particular, owing to their extraordinary thermal conductivity and mechanical and electrical properties, carbon nanotubes find applications as additives to various structural materials. For instance, nanotubes form a tiny portion of the material(s) in some (primarily carbon fibre) baseball bats, golf clubs, or car parts.

Nanotubes are members of the fullerene structural family. Their name is derived from their long, hollow structure with the walls formed by one-atom-thick sheets of carbon, called graphene. These sheets are rolled at specific and discrete ("chiral") angles, and the combination of the rolling angle and radius decides the properties; for example, whether the individual nanotube shell is a metal or semiconductor. Nanotubes are categorized as single-walled nanotubes (SWNTs) and multi-walled nanotubes (MWNTs). Individual nanotubes naturally align themselves into "ropes" held together by van der Waals forces, more specifically, pi-stacking.

Applied quantum chemistry, specifically, orbital hybridization best describes chemical bonding in nanotubes. The chemical bonding of nanotubes is composed entirely of sp^2 bonds, similar to those of graphite. These bonds, which are stronger than the sp^3 bonds found in alkanes and diamond, provide nanotubes with their unique strength.

Mechanical Properties of Carbon Nanotubes

Carbon nanotube is one of the strongest materials in nature. Carbon nanotubes (CNTs) are basically long hollow cylinders of graphite sheets. Although a graphite sheet has a 2D symmetry, carbon nanotubes by geometry have different properties in axial and radial directions. It has been shown that CNTs are very strong in the axial direction. Young's modulus on the order of 270 - 950 GPa and tensile strength of 11 - 63 GPa were obtained.

On the other hand, there was evidence that in the radial direction they are rather soft. The first transmission electron microscope observation of radial elasticity suggested that even the van der Waals forces can deform two adjacent nanotubes. Later, nanoindentations with atomic force microscope were performed by several groups to quantitatively measure radial elasticity of multiwalled carbon nanotubes and tapping/contact mode atomic force microscopy was recently performed on single-walled carbon nanotubes. Young's modulus of on the order of several GPa showed that CNTs are in fact very soft in the radial direction. Radial direction elasticity of CNTs is important especially for carbon nanotube composites where the embedded tubes are subjected to large deformation in the transverse direction under the applied load on the composite structure. One of the main problems in characterizing the radial elasticity of CNTs is the knowledge about the internal radius of the CNT; carbon nanotubes with identical outer diameter may have different internal diameter (or the number of walls). Recently a method using an atomic force microscope was introduced to find the exact number of layers and hence the internal diameter of the CNT. In this way, mechanical characterization is more accurate.

Optical Properties of Carbon Nanotubes

Within materials science, the optical properties of carbon nanotubes refer specifically to the absorption, photoluminescence (fluorescence), and Raman spectroscopy of carbon nanotubes. Spectroscopic methods offer the possibility of quick and non-destructive characterization of relatively large amounts of carbon nanotubes. There is a strong demand for such characterization from the industrial point of view: numerous parameters of the nanotube synthesis can be changed, intentionally or unintentionally, to alter the nanotube quality. As shown below, optical absorption, photoluminescence and Raman spectroscopies allow quick and reliable characterization of this "nanotube quality" in terms of non-tubular carbon content, structure (chirality) of the produced nanotubes, and structural defects. Those features determine nearly any other properties such as optical, mechanical, and electrical properties.

Carbon nanotubes are unique "one-dimensional systems" which can be envisioned as rolled single sheets of graphite (or more precisely graphene). This rolling can be done at different angles and curvatures resulting in different nanotube properties. The diameter typically varies in the range 0.4–40 nm (i.e. "only" ~100 times), but the length can vary ~10,000 times, reaching 18.5 cm. Thus the nanotube aspect ratio, or the length-to-diameter ratio, can be as high as 132,000,000:1, which is unequalled by any other material. Consequently, all the properties of the carbon nanotubes relative to those of typical semiconductors are extremely anisotropic (directionally dependent) and tunable.

Whereas mechanical, electrical and electrochemical (supercapacitor) properties of the carbon nanotubes are well established and have immediate applications, the practical use of optical properties is yet unclear. The aforementioned tunability of properties is potentially useful in optics and photonics. In particular, light-emitting diodes (LEDs) and photo-detectors based on a single nanotube have been produced in the lab. Their unique feature is not the efficiency, which is yet relatively low, but the narrow selectivity in the wavelength of emission and detection of light and the possibility of its fine tuning through the nanotube structure. In addition, bolometer and optoelectronic memory devices have been realised on ensembles of single-walled carbon nanotubes.

Terminology

This article uses the following abbreviations:

- Carbon nanotube (CNT)
- Single wall carbon nanotube (SWCNT)
- Multiwall carbon nanotube (MWCNT)

However, C is often omitted in scientific literature, so NT, SWNT and MWNT are more commonly used. Also, "wall" is often exchanged with "walled".

Electronic Structure of Carbon Nanotube

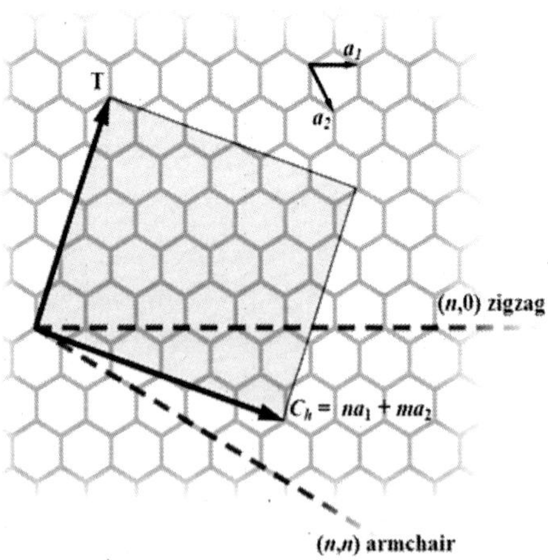

Figure: *The* (n, m) *nanotube naming scheme can be thought of as a vector* (C_h) *in an infinite graphene sheet that describes how to "roll up" the graphene sheet to make the nanotube. T denotes the tube axis, and* a_1 *and* a_2 *are the unit vectors of graphene in real space.*

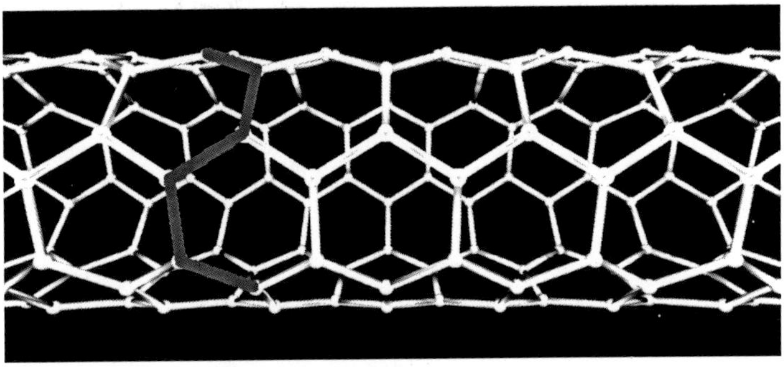

Figure: *Armchair nanotube*

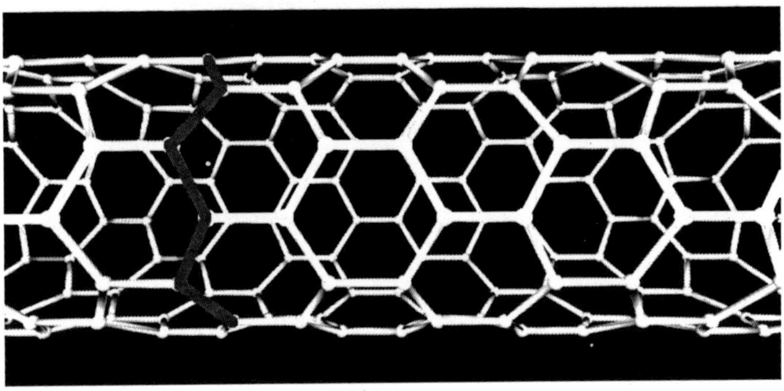

Figure: *Zigzag nanotube*

A single-wall carbon nanotube can be imagined as graphene sheet rolled at a certain "chiral" angle with respect to a plane perpendicular to the tube's long axis. Consequently, SWCNT can be defined by its diameter and chiral angle. The chiral angle can range from 0 to 30 degrees.

However, more conveniently, a pair of indices (n, m) is used instead. The indices refer to equally long unit vectors at 60° angles to each other across a single 6-member carbon ring. Taking the origin as carbon number 1, the a_1 unit vector may be considered the line drawn from carbon 1 to carbon 3, and the a_2 unit vector is then the line drawn from carbon 1 to carbon 5. To visualize a CNT with indices (n, m), draw n a_1 unit vectors across the graphene sheet, then draw m a_2 unit vectors at a 60° angle to the a_1 vectors, then add the vectors together. The line representing the sum of the vectors will define the circumference of the CNT along the plane perpendicular to its long axis, connecting one end to the other. In the diagram at right, C_h is a (4, 2) vector: the sum of 4 unit vectors from the origin directly to the right, then 2 unit vectors at a 60° angle down and to the right.

Tubes having $n = m$ (chiral angle = 30°) are called "armchair" and those with $m = 0$ (chiral angle = 0°) "zigzag". Those indices uniquely determine whether CNT is a metal, semimetal or semiconductor, as well as its band gap: when $|m - n| = 3k$ (k is integer), the tube is metallic; but if $|m - n| = 3k \pm 1$, the tube is semiconducting. The nanotube diameter d is related to m and n as

$$d = \frac{a}{\pi}\sqrt{(n^2 + nm + m^2)}.$$

In this equation, a = 0.246 nm is the magnitude of either unit vector a_1 or a_2.

The situation in multi-wall CNTs is complicated as their properties are determined by contribution of all individual shells; those shells have different structures, and, because of the synthesis, are usually more defective than SWCNTs. Therefore, optical properties of MWCNTs will not be considered here.

Van Hove Singularities

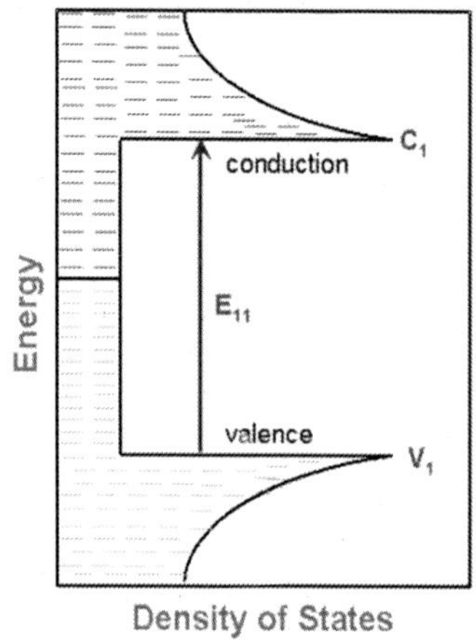

Metallic SWNT

$v_1 \rightarrow c_1$ corresponds to the "first van Hove" optical transition

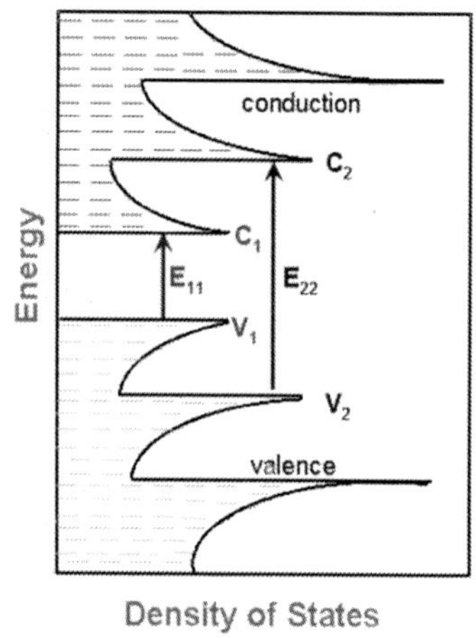

Semiconducting SWNT

$v_2 \rightarrow c_2$ corresponds to the "second van Hove" optical transition

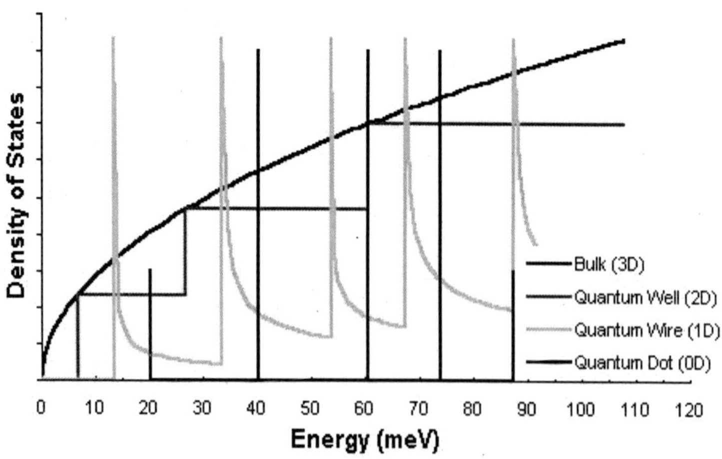

Figure: *A bulk 3D material (blue) has continuous DOS, but a 1D wire (green) has Van Hove singularities.*

Optical properties of carbon nanotubes derive from electronic transitions within one-dimensional density of states (DOS). A typical feature of one-dimensional crystals is that their DOS is not a continuous function of energy, but it descends gradually and then increases in a discontinuous spike. In contrast, three-dimensional materials have continuous DOS. The sharp peaks found in one-dimensional materials are called Van Hove singularities.

Van Hove singularities result in the following remarkable optical properties of carbon nanotubes:

- Optical transitions occur between the $v_1 - c_1$, $v_2 - c_2$, etc., states of semiconducting or metallic nanotubes and are traditionally labelled as S_{11}, S_{22}, M_{11}, etc., or, if the "conductivity" of the tube is unknown or unimportant, as E_{11}, E_{22}, etc. Crossover transitions $c_1 - v_2$, $c_2 - v_1$, etc., are dipole-forbidden and thus are extremely weak, but they were possibly observed using cross-polarized optical geometry.
- The energies between the Van Hove singularities depend on the nanotube structure. Thus by varying this structure, one can tune the optoelectronic properties of carbon nanotube. Such fine tuning has been experimentally demonstrated using UV illumination of polymer-dispersed CNTs.
- Optical transitions are rather sharp (~10 meV) and strong. Consequently, it is relatively easy to selectively excite nanotubes having certain (n, m) indices, as well as to detect optical signals from individual nanotubes.

Kataura Plot

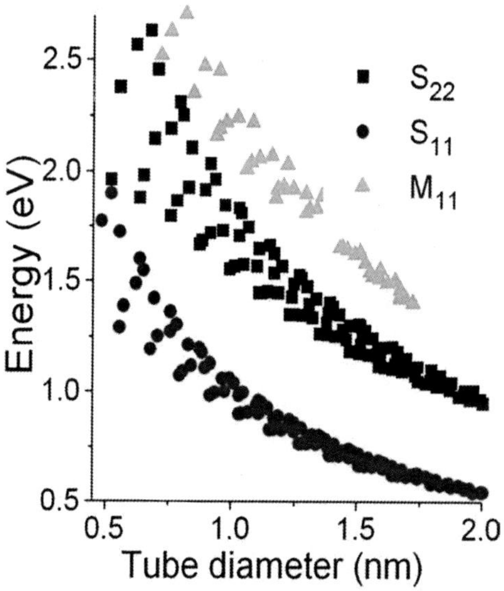

Figure: *In this Kataura plot, the energy of an electronic transition decreases as the diameter of the nanotube increases.*

The band structure of carbon nanotubes having certain (*n*, *m*) indexes can be easily calculated. A theoretical graph based on this calculations was designed in 1999 by Hiromichi Kataura to rationalize experimental findings. A Kataura plot relates the nanotube diameter and its bandgap energies for all nanotubes in a diameter range. The oscillating shape of every branch of the Kataura plot reflects the intrinsic strong dependence of the SWCNT properties on the (*n*, *m*) index rather than on its diameter. For example, (10, 1) and (8, 3) tubes have almost the same diameter, but very different properties: the former is a metal, but the latter is semiconductor.

Optical Absorption

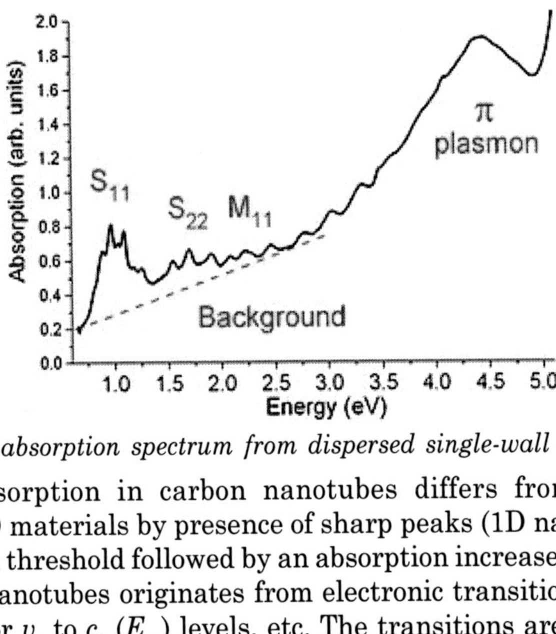

***Figure:** Optical absorption spectrum from dispersed single-wall carbon nanotubes*

Optical absorption in carbon nanotubes differs from absorption in conventional 3D materials by presence of sharp peaks (1D nanotubes) instead of an absorption threshold followed by an absorption increase (most 3D solids). Absorption in nanotubes originates from electronic transitions from the v_2 to c_2 (energy E_{22}) or v_1 to c_1 (E_{11}) levels, etc. The transitions are relatively sharp and can be used to identify nanotube types. Note that the sharpness deteriorates with increasing energy, and that many nanotubes have very similar E_{22} or E_{11} energies, and thus significant overlap occurs in absorption spectra. This overlap is avoided in photoluminescence mapping measurements, which instead of a combination of overlapped transitions identifies individual (E_{22}, E_{11}) pairs.

Interactions between nanotubes, such as bundling, broaden optical lines. While bundling strongly affects photoluminescence, it has much weaker effect on optical absorption and Raman scattering. Consequently, sample preparation for the latter two techniques is relatively simple.

Optical absorption is routinely used to quantify quality of the carbon nanotube powders.

The spectrum is analyzed in terms of intensities of nanotube-related peaks, background and pi-carbon peak; the latter two mostly originate from non-nanotube carbon in contaminated samples. However, it has been recently shown that by aggregating nearly single chirality semiconducting nanotubes into closely packed Van der Waals bundles the absorption background can be attributed to free carrier transition originating from intertube charge transfer.

Carbon Nanotubes as a Black Body

An ideal black body should have emissivity or absorbance of 1.0, which is difficult to attain in practice, especially in a wide spectral range. Vertically aligned "forests" of single-wall carbon nanotubes can have absorbances of 0.98–0.99 from the far-ultraviolet (200 nm) to far-infrared (200 μm) wavelengths. Super black, a coating based on chemically etched nickel-phosphorus alloy, is another material approaching the absorption of 1.0.

These SWNT forests (buckypaper) were grown by the super-growth CVD method to about 10 μm height. Two factors could contribute to strong light absorption by these structures: (i) a distribution of CNT chiralities resulted in various bandgaps for individual CNTs. Thus a compound material was formed with broadband absorption. (ii) Light might be trapped in those forests due to multiple reflections.

Reflectance measurements			
	UV-to-near IR	*Near-to-mid IR*	*Mid-to-far IR*
Wavelength, μm	0.2-2	2–20	25–200
Incident angle, °	8	5	10
Reflection	Hemispherical-directional	Hemispherical-directional	Specular
Reference	White reflectance standard	Gold mirror	Aluminum mirror
Average reflectance	0.0160	0.0097	0.0017
Standard deviation	0.0048	0.0041	0.0027

Luminescence

Photoluminescence (Fluorescence): Semiconducting single-walled carbon nanotubes emit near-infrared light upon photoexcitation, described interchangeably as fluorescence or photoluminescence (PL). The excitation of PL usually occurs as follows: an electron in a nanotube absorbs excitation light via S_{22} transition, creating an electron-hole pair (exciton). Both electron and hole rapidly relax (via phonon-assisted processes) from c_2 to c_1 and from v_2 to v_1 states, respectively. Then they recombine through a $c_1 - v_1$ transition resulting in light emission.

No excitonic luminescence can be produced in metallic tubes. Their electrons can be excited, thus resulting in optical absorption, but the holes are immediately filled by other electrons out of the many available in the metal. Therefore, no excitons are produced.

Salient Properties

- Photoluminescence from SWCNT, as well as optical absorption and Raman scattering, is linearly polarized along the tube axis. This allows monitoring of the SWCNTs orientation without direct microscopic observation.
- PL is quick: relaxation typically occurs within 100 picoseconds.
- PL efficiency was first found to be low (~0.01%), but later studies measured much higher quantum yields. By improving the structural quality and isolation of nanotubes, emission efficiency increased. A quantum yield of 1% was reported in nanotubes sorted by diameter and length through gradient centrifugation, and it was further increased to 20% by optimizing the procedure of isolating individual nanotubes in solution.
- The spectral range of PL is rather wide. Emission wavelength can vary between 0.8 and 2.1 micrometres depending on the nanotube structure.
- Excitons are apparently delocalized over several nanotubes in single chirality bundles as the photoluminescence spectrum displays a splitting consistent with intertube exciton tunneling.
- Interaction between nanotubes or between a nanotube and another material may quench or increase PL. No PL is observed in multi-walled carbon nanotubes. PL from double-wall carbon nanotubes strongly depends on the preparation method: CVD grown DWCNTs show emission both from inner and outer shells. However, DWCNTs produced by encapsulating fullerenes into SWCNTs and annealing show PL only from the outer shells. Isolated SWCNTs lying on the substrate show extremely weak PL which has been detected in few studies only. Detachment of the tubes from the substrate drastically increases PL.
- Position of the (S_{22}, S_{11}) PL peaks depends slightly (within 2%) on the nanotube environment (air, dispersant, etc.). However, the shift depends on the (*n*, *m*) index, and thus the whole PL map not only shifts, but also warps upon changing the CNT medium.

Applications

- Photoluminescence is used for characterization purposes to measure the quantities of semiconducting nanotube species in a sample. Nanotubes are isolated (dispersed) using an appropriate chemical agent ("dispersant") to reduce the intertube quenching. Then PL is measured, scanning both the excitation and emission energies and thereby producing a PL map. The ovals in the map define (S_{22}, S_{11}) pairs, which unique identify (*n*, *m*) index of a tube. The data of Weisman and Bachilo are conventionally used for the identification.

- Nanotube fluorescence has been investigated for the purposes of imaging and sensing in biomedical applications.

Sensitization

Optical properties, including the PL efficiency, can be modified by encapsulating organic dyes (carotene, lycopene, etc.) inside the tubes. Efficient energy transfer occurs between the encapsulated dye and nanotube — light is efficiently absorbed by the dye and without significant loss is transferred to the SWCNT. Thus potentially, optical properties of a carbon nanotube can be controlled by encapsulating certain molecule inside it. Besides, encapsulation allows isolation and characterization of organic molecules which are unstable under ambient conditions. For example, Raman spectra are extremely difficult to measure from dyes because of their strong PL (efficiency close to 100%). However, encapsulation of dye molecules inside SWCNTs completely quenches dye PL, thus allowing measurement and analysis of their Raman spectra.

Cathodoluminescence

Cathodoluminescence (CL) — light emission excited by electron beam — is a process commonly observed in TV screens. An electron beam can be finely focused and scanned across the studied material. This technique is widely used to study defects in semiconductors and nanostructures with nanometre-scale spatial resolution. It would be beneficial to apply this technique to carbon nanotubes. However, no reliable CL, i.e. sharp peaks assignable to certain (*n*, *m*) indices, has been detected from carbon nanotubes yet.

Electroluminescence

If appropriate electrical contacts are attached to a nanotube, electron-hole pairs (excitons) can be generated by injecting electrons and holes from the contacts. Subsequent exciton recombination results in electroluminescence (EL). Electroluminescent devices have been produced from single nanotubes.

Raman Scattering

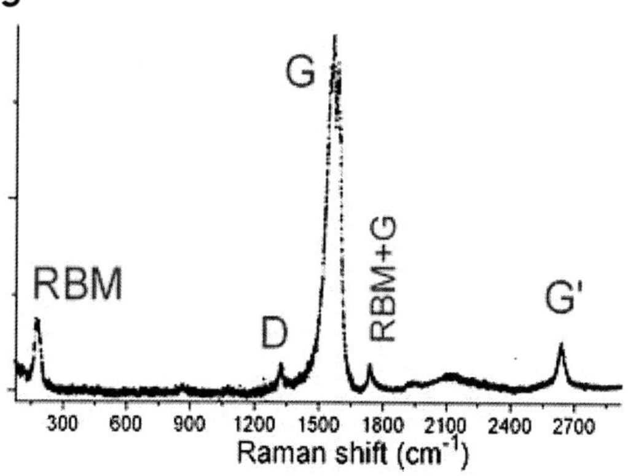

Figure: *Raman spectrum of single-wall carbon nanotubes*

Raman spectroscopy has good spatial resolution (~0.5 micrometres) and sensitivity (single nanotubes); it requires only minimal sample preparation and is rather informative. Consequently, Raman spectroscopy is probably the most popular technique of carbon nanotube characterization. Raman scattering in SWCNTs is resonant, i.e., only those tubes are probed which have one of the bandgaps equal to the exciting laser energy. Several scattering modes dominate the SWCNT spectrum, as discussed below.

Similar to photoluminescence mapping, the energy of the excitation light can be scanned in Raman measurements, thus producing Raman maps. Those maps also contain oval-shaped features uniquely identifying (*n*, *m*) indices. Contrary to PL, Raman mapping detects not only semiconducting but also metallic tubes, and it is less sensitive to nanotube bundling than PL. However, requirement of a tunable laser and a dedicated spectrometre is a strong technical impediment.

Radial Breathing Mode

Radial breathing mode (RBM) corresponds to radial expansion-contraction of the nanotube. Therefore, its frequency n_{RBM} (in cm^{-1}) depends on the nanotube diameter d as, $n_{RBM} = A/d + B$ (where A and B are constants dependent on the environment in which the nanotube is present. For example B=0 for individual nanotubes.) (in nanometres) and can be estimated as $n_{RBM} = 234/d + 10$ for SWNT or $n_{RBM} = 248/d$ for DWNT, which is very useful in deducing the CNT diameter from the RBM position. Typical RBM range is 100–350 cm^{-1}. If RBM intensity is particularly strong, its weak second overtone can be observed at double frequency.

Bundling Mode

The bundling mode is a special form of RBM supposedly originating from collective vibration in a bundle of SWCNTs.

G Mode

Another very important mode is the G mode (G from graphite). This mode corresponds to planar vibrations of carbon atoms and is present in most graphite-like materials. G band in SWCNT is shifted to lower frequencies relative to graphite (1580 cm^{-1}) and is split into several peaks. The splitting pattern and intensity depend on the tube structure and excitation energy; they can be used, though with much lower accuracy compared to RBM mode, to estimate the tube diameter and whether the tube is metallic or semiconducting. A wider pattern indicates semiconducting nature of the nanotube.

D Mode

D mode is present in all graphite-like carbons and originates from structural defects. Therefore, the ratio of the *G*/*D* modes is conventionally used to quantify the structural quality of carbon nanotubes. High-quality nanotubes have this ratio significantly higher than 100. At a lower functionalisation of the nanotube, the *G/D* ratio remains almost unchanged. This ratio gives an idea of the functionalisation of a nanotube.

G' Mode

The name of this mode is misleading: it is given because in graphite, this mode is usually the second strongest after the G mode. However, it is actually the second overtone of the defect-induced D mode (and thus should logically be named D'). Its intensity is stronger than that of the D mode due to different selection rules. In particular, D mode is forbidden in the ideal nanotube and requires a structural defect, providing a phonon of certain angular momentum, to be induced. In contrast, G' mode involves a "self-annihilating" pair of phonons and thus does not require defects. The spectral position of G' mode depends on diameter, so it can be used roughly to estimate the SWCNT diameter. In particular, G' mode is a doublet in double-wall carbon nanotubes, but the doublet is often unresolved due to line broadening.

Other overtones, such as a combination of RBM+G mode at ~1750 cm^{-1}, are frequently seen in CNT Raman spectra. However, they are less important and are not considered here.

Anti-Stokes Scattering

All the above Raman modes can be observed both as Stokes and anti-Stokes scattering. As mentioned above, Raman scattering from CNTs is resonant in nature, i.e. only tubes whose band gap energy is similar to the laser energy are excited. The difference between those two energies, and thus the band gap of individual tubes, can be estimated from the intensity ratio of the Stokes/anti-Stokes lines. This estimate however relies on the temperature factor (Boltzmann factor), which is often miscalculated – focused laser beam is used in the measurement, which can locally heat the nanotubes without changing the overall temperature of the studied sample.

Rayleigh Scattering

Carbon nanotubes have very large aspect ratio, i.e., their length is much larger than their diameter. Consequently, as expected from the classical electromagnetic theory, elastic light scattering (or Rayleigh scattering) by straight CNTs has anisotropic angular dependence, and from its spectrum, the band gaps of individual nanotubes can be deduced.

Another manifestation of Rayleigh scattering is the "antenna effect", an array of nanotubes standing on a substrate has specific angular and spectral distributions of reflected light, and both those distributions depend on the nanotube length.

Boron Nitride Nanotube

Boron nitride is a chemical compound with chemical formula BN, consisting of equal numbers of boron and nitrogen atoms. BN is isoelectronic to a similarly structured carbon lattice and thus exists in various crystalline forms. The hexagonal form corresponding to graphite is the most stable and softest among BN polymorphs, and is therefore used as a lubricant and an additive to cosmetic products. The cubic (sphalerite structure) variety analogous to diamond is called c-BN. Its hardness is inferior only to diamond, but its thermal and chemical stability is superior. The rare wurtzite BN modification is similar to lonsdaleite and may even be harder than the cubic form.

Boron nitride is rarely found in nature, as dispersed micrometre-sized inclusions of qingsongite (c-BN) in chromium-rich rocks in Tibet. It is therefore produced synthetically from boric acid or boron trioxide. The initial product is amorphous BN powder, which is converted to crystalline h-BN by heating in nitrogen flow at temperatures above 1500 °C. c-BN is made by annealing h-BN powder at higher temperatures, under pressures above 5 GPa. Contrary to diamond, larger c-BN pellets can be produced by fusing (sintering) relatively cheap c-BN powders. As a result, c-BN is widely used in mechanical applications.

Because of excellent thermal and chemical stability, boron nitride ceramics are traditionally used as parts of high-temperature equipment. Boron nitride has a great potential in nanotechnology. Nanotubes of BN can be produced that have a structure similar to that of carbon nanotubes, i.e. graphene (or BN) sheets rolled on themselves, but the properties are very different: whereas carbon nanotubes can be metallic or semiconducting depending on the rolling direction and radius, a BN nanotube is an electrical insulator with a wide bandgap of ~5.5 eV (same as in diamond), which is almost independent of tube chirality and morphology. Similar to other BN forms, BN nanotubes are more thermally and chemically stable than carbon nanotubes, which favours them for some applications.

Structure

Boron nitride has been produced in an amorphous (a-BN) and crystalline forms. The most stable crystalline form is the hexagonal one, also called h-BN, α-BN, or g-BN (graphitic BN). It has a layered structure similar to graphite. Within each layer, boron and nitrogen atoms are bound by strong covalent bonds, whereas the layers are held together by weak van der Waals forces. The interlayer "registry" of these sheets differs, however, from the pattern seen for graphite, because the atoms are eclipsed, with boron atoms lying over and above nitrogen atoms. This registry reflects the polarity of the B-N bonds. Still, h-BN and graphite are very close neighbours and even the BC_6N hybrids have been synthesized where carbon substitutes for some B and N atoms.

As diamond is less stable than graphite, cubic BN is less stable than h-BN, but the conversion rate between those forms is negligible at room temperature (again like diamond). The cubic form has the sphalerite crystal structure, the same as that of diamond, and is also called β-BN or c-BN. The wurtzite BN form (w-BN) has the same structure as lonsdaleite, a rare hexagonal polymorph of carbon. In both c-BN and w-BN boron and nitrogen atoms are grouped into tetrahedra, but the angles between neighbouring tetrahedra are different.

Properties

Physical: Properties of amorphous and crystalline BN, graphite and diamond. Some properties of h-BN and graphite differ within the basal planes (Q%) and perpendicular to them (g%)

Properties of amorphous and crystalline BN, graphite and diamond. Some properties of h-BN and graphite differ within the basal planes (‖) and perpendicular to them (⊥)

Material	a-BN	h-BN	c-BN	w-BN	graphite	diamond
Density (g/cm³)	2.28	~2.1	3.45	3.49	~2.1	3.515
Mohs hardness		1–2	9.5–10	~10	1–2	10
Knoop hardness (GPa)	10		45	34		100
Bulk modulus (GPa)	100	36.5	400	400	34	440
Thermal conductivity (W/m.K)	3	600‖; 30 ⊥	740		200–2000‖; 2–800⊥	600–2000
Thermal expansion (10^{-6}/°C)		–2.7‖; 38⊥	1.2	2.7	–1.5‖; 25⊥	0.8
Bandgap (eV)	5.05	5.2	6.4	4.5–5.5	0	5.5
Refractive index	1.7	1.8	2.1	2.05		2.4
Magnetic susceptibility (μemu/g)[4]		–0.48‖; -17.3⊥			–0.2..-2.7‖; -20..-28⊥	–1.6

Sources: amorphous BN, crystalline BN, graphite, diamond.

The partly ionic structure of BN layers in h-BN reduces covalency and electrical conductivity, whereas the interlayer interaction increases resulting in higher hardness of h-BN relative to graphite. The reduced electron-delocalization in hexagonal-BN is also indicated by its absence of colour and a large band gap. Very different bonding – strong covalent within the basal planes (planes where boron and nitrogen atoms are covalently bonded) and weak between them – causes high anisotropy of most properties of h-BN.

For example, the hardness, electrical and thermal conductivity are much higher within the planes than perpendicular to them. On the contrary, the properties of c-BN and w-BN are more homogeneous and isotropic.

Those materials are extremely hard, with the hardness of bulk c-BN being slightly smaller and w-BN even higher than that of diamond. Polycrystalline c-BN with grain sizes on the order of 10 nm is also reported to have Vickers hardness comparable or higher than diamond. Because of much better stability to heat and metals, c-BN surpasses diamond in mechanical applications. The thermal conductivity of BN is among the highest of all electric insulators.

Boron nitride can be doped p-type with Be and n-type with boron, sulfur, silicon or if co-doped with carbon and nitrogen. Both hexagonal and cubic BN are wide-gap semiconductors with a band gap energy corresponding to the UV region. If voltage is applied to h-BN or c-BN, then it emits UV light in the range 215–250 nm and therefore can potentially be used as light emitting diodes (LEDs) or lasers.

Little is known on melting behaviour of boron nitride. It sublimates at 2973 °C at normal pressure releasing nitrogen gas and boron, but melts at elevated pressure.

Thermal Stability

Hexagonal and cubic (and probably w-BN) BN show remarkable chemical and thermal stabilities. For example, h-BN is stable to decomposition at temperatures up to 1000 °C in air, 1400 °C in vacuum, and 2800 °C in an inert atmosphere. The reactivity of h-BN and c-BN is relatively similar, and the data for c-BN are summarized in the table below.

Reactivity of c-BN with solids			
Solid	***Ambient***	***Action***	***Threshold T (°C)***
Mo	10^{-2} Pa vacuum	reaction	1360
Ni	10^{-2} Pa vacuum	wetting	1360
Fe, Ni, Co	argon	react	1400–1500
Al	10^{-2} Pa vacuum	wetting and reaction	1050
Si	10^{-3} Pa vacuum	wetting	1500
Cu, Ag, Au, Ga, In, Ge, Sn	10^{-3} Pa vacuum	no wetting	1100
B		no wetting	2200
$Al_2O_3 + B_2O_3$	10^{-2} Pa vacuum	no reaction	1360

Thermal stability of c-BN can be summarized as follows:

- In air or oxygen: B_2O_3 protective layer prevents further oxidation to ~1300 °C; no conversion to hexagonal form at 1400 °C.
- In nitrogen: some conversion to h-BN at 1525 °C after 12 h.
- In vacuum (10^{-5} Pa): conversion to h-BN at 1550–1600 °C.

Chemical Stability

Boron nitride is insoluble in usual acids, but is soluble in alkaline molten salts and nitrides, such as LiOH, KOH, NaOH-Na_2CO_3, $NaNO_3$, Li_3N, Mg_3N_2, Sr_3N_2, Ba_3N_2 or Li_3BN_2, which are therefore used to etch BN.

Thermal Conductivity

The theoretical thermal conductivity of hexagonal Boron nitride nanoribbons (BNNRs) can approach 1700–2000 W/(m·K), which has the same order of magnitude as the experimental measured value for graphene, and can be comparable to the theoretical calculations for graphene nanoribbons. Moreover, the thermal transport in the BNNRs is anisotropic. The thermal conductivity of zigzag-edged BNNRs is about 20% larger than that of armchair-edged nanoribbons at room temperature.

Synthesis

Preparation and Reactivity of Hexagonal BN: Boron nitride has not been found in nature and therefore is produced synthetically. Hexagonal boron nitride is obtained by the reacting boron trioxide (B_2O_3) or boric acid ($B(OH)_3$) with ammonia (NH_3) or urea ($CO(NH_2)_2$) in a nitrogen atmosphere:

$$B_2O_3 + 2\ NH_3 \rightarrow 2\ BN + 3\ H_2O\ (T = 900\ °C)$$

$$B(OH)_3 + NH_3 \rightarrow BN + 3\ H_2O\ (T = 900\ °C)$$

$$B_2O_3 + CO(NH_2)_2 \rightarrow 2\ BN + CO_2 + 2\ H_2O\ (T > 1000\ °C)$$

$$B_2O_3 + 3\ CaB_6 + 10\ N_2 \rightarrow 20\ BN + 3\ CaO\ (T > 1500\ °C)$$

The resulting disordered (amorphous) boron nitride contains 92–95% BN and 5–8% B_2O_3. The remaining B_2O_3 can be evaporated in a second step at temperatures > 1500 °C in order to achieve BN concentration >98%. Such annealing also crystallizes BN, the size of the crystallites increasing with the annealing temperature.

h-BN parts can be fabricated inexpensively by hot-pressing with subsequent machining. The parts are made from boron nitride powders adding boron oxide for better compressibility. Thin films of boron nitride can be obtained by chemical vapor deposition from boron trichloride and nitrogen precursors. Combustion of boron powder in nitrogen plasma at 5500 °C yields ultrafine boron nitride used for lubricants and toners.

Boron nitride reacts with iodine fluoride in trichlorofluoromethane at "30 °C to produce an extremely sensitive contact explosive, NI_3, in low yield. Boron nitride reacts with nitrides of alkali metals and lanthanides to form nitridoborate compounds. For example:

$$Li_3N + BN \rightarrow Li_3BN_2$$

Intercalation of Hexagonal BN

Similar to graphite, various molecules, such as NH_3 or alkali metals, can be intercalated into hexagonal boron nitride, that is inserted between its layers. Both experiment and theory suggest the intercalation is much more difficult for BN than for graphite.

Preparation of Cubic BN

Synthesis of c-BN uses same methods as that of diamond: Cubic boron nitride is produced by treating hexagonal boron nitride at high pressure and temperature, much as synthetic diamond is produced from graphite. Direct conversion of hexagonal boron nitride to the cubic form has been observed at pressures between 5 and 18 GPa and temperatures between 1730 and 3230 °C, that is similar parameters as for direct graphite-diamond conversion. The addition of a small amount of boron oxide can lower the required pressure to 4–7 GPa and temperature to 1500 °C. As in diamond synthesis, to further reduce the conversion pressures and temperatures, a catalyst is added, such as lithium, potassium, or magnesium, their nitrides, their fluoronitrides, water with ammonium compounds, or hydrazine. Other industrial synthesis methods, again borrowed from diamond growth, use crystal growth in a temperature gradient, or explosive shock wave. The shock wave method is used to produce material called heterodiamond, a superhard compound of boron, carbon, and nitrogen. Low-pressure deposition of thin films of cubic boron nitride is possible. As in diamond growth, the major problem is to suppress the growth of hexagonal phases (h-BN or graphite, respectively). Whereas in diamond growth this is achieved by adding hydrogen gas, boron trifluoride is used for c-BN. Ion beam deposition, plasma-enhanced chemical vapor deposition, pulsed laser deposition, reactive sputtering, and other physical vapor deposition methods are used as well.

Preparation of Wurtzite BN

Wurtzite BN can be obtained via static high-pressure or dynamic shock methods. The limits of its stability are not well defined. Both c-BN and w-BN are formed by compressing h-BN, but formation of w-BN occurs at much lower temperatures close to 1700 °C.

Production Statistics

Whereas the production and consumption figures for the raw materials used for BN synthesis, namely boric acid and boron trioxide, are well known, the corresponding numbers for the boron nitride are not listed in statistical reports. An estimate for the 1999 world production is 300 to 350 metric tons. The major producers and consumers of BN are located in the United States, Japan, China and Germany. In 2000, prices varied from about $75/kg to $120/kg for standard industrial-quality h-BN and were about up to $200–$400/kg for high purity BN grades.

Applications

Hexagonal BN: Hexagonal BN is the most widely used polymorph. It is a good lubricant at both low and high temperatures (up to 900 °C, even in an oxidizing atmosphere). h-BN lubricant is particularly useful when the electrical conductivity or chemical reactivity of graphite (alternative lubricant) would be problematic. Another advantage of h-BN over graphite is that its lubricity does not require water or gas molecules trapped between the layers. Therefore, h-BN lubricants can be used even in vacuum, e.g. in space applications. The lubricating properties of fine-grained h-BN are used in cosmetics, paints, dental cements, and pencil leads.

Hexagonal BN was first used in cosmetics around 1940 in Japan. However, because of its high price, h-BN was soon abandoned for this application. Its use was revitalized in the late 1990s with the optimization h-BN production processes, and currently h-BN is used by nearly all leading producers of cosmetic products for foundations, make-up, eye shadows, blushers, kohl pencils, lipsticks and other skincare products.

Because of its excellent thermal and chemical stability, boron nitride ceramics are traditionally used as parts of high-temperature equipment. h-BN can be included in ceramics, alloys, resins, plastics, rubbers, and other materials, giving them self-lubricating properties. Such materials are suitable for construction of e.g. bearings and in steelmaking. Plastics filled with BN have less thermal expansion as well as higher thermal conductivity and electrical resistivity. Due to its excellent dielectric and thermal properties, BN is used in electronics e.g. as a substrate for semiconductors, microwave-transparent windows, and as a structural material for seals.

Hexagonal BN is used in xerographic process and laser printers as a charge leakage barrier layer of the photo drum. In the automotive industry, h-BN mixed with a binder (boron oxide) is used for sealing oxygen sensors, which provide feedback for adjusting fuel flow. The binder utilises the unique temperature stability and insulating properties of h-BN.

Parts can be made of h-BN by hot pressing. Union Carbide Corporation produces three grades of BN. HBN, with boron oxide binder, usable to 550–850 °C in oxidizing atmosphere and up to 1600 °C in vacuum, but due to the boron oxide content is sensitive to water. HBR uses calcium borate binder and is usable at 1600 °C. HBC grade uses no binder and can be used to 3000 °C.

Cubic Boron Nitride

Cubic boron nitride (CBN or c-BN) is widely used as an abrasive. Its usefulness arises from its insolubility in iron, nickel, and related alloys at high temperatures, whereas diamond is soluble in these metals to give carbides. Polycrystalline c-BN (PCBN) abrasives are therefore used for machining steel, whereas diamond abrasives are preferred for aluminum alloys, ceramics, and stone. When in contact with oxygen at high temperatures, BN forms a passivation layer of boron oxide. Boron nitride binds well with metals, due to formation of interlayers of metal borides or nitrides. Materials with cubic boron nitride crystals are often used in the tool bits of cutting tools. For grinding applications, softer binders, e.g. resin, porous ceramics, and soft metals, are used. Ceramic binders can be used as well. Commercial products are known under names "Borazon" (by Diamond Innovations), and "Elbor" or "Cubonite" (by Russian vendors). Similar to diamond, the combination in c-BN of highest thermal conductivity and electrical resistivity is ideal for heat spreaders. Contrary to diamond, large c-BN pellets can be produced in a simple process (called sintering) of annealing c-BN powders in nitrogen flow at temperatures slightly below the BN decomposition temperature. This ability of c-BN and h-BN powders to fuse allows cheap production of large BN parts.

As cubic boron nitride consists of light atoms and is very robust chemically and mechanically, it is one of the popular materials for X-ray membranes: low mass results in small X-ray absorption, and good mechanical properties allow usage of thin membranes, thus further reducing the absorption.

Amorphous Boron Nitride

Layers of amorphous boron nitride (a-BN) are used in some semiconductor devices, e.g. MISFETs. They can be prepared by chemical decomposition of trichloroborazine with caesium, or by thermal chemical vapor deposition methods. Thermal CVD can be also used for deposition of h-BN layers, or at high temperatures, c-BN.

Buckypaper

Buckypaper is a thin sheet made from an aggregate of carbon nanotubes. The nanotubes are approximately 50,000 times thinner than a human hair. Originally, it was fabricated as a way to handle carbon nanotubes, but it is also being studied and developed into applications by several research groups, showing promise as vehicle armor, personal armor, and next-generation electronics and displays.

The generally accepted methods of making CNT films involves the use of non-ionic surfactants, such as Triton X-100 and sodium lauryl sulfate, which improves their dispersibility in aqueous solution. These suspensions can then be membrane filtered under positive or negative pressure to yield uniform films. The van der Waals force's interaction between the nanotube surface and the surfactant can often be mechanically strong and quite stable and therefore there are no assurances that all the surfactant is removed from the CNT film after formation. Washing with methanol, an effective solvent in the removal of Triton X, was found to cause cracking and deformation of the film. It has also been found that Triton X can lead to cell lysis and in turn tissue inflammatory responses even at low concentrations.

In order to avoid adverse side-effects from the possible presence of surfactants, an alternative casting process can be used involving a frit compression method that did not require the use of surfactants or surface modification. The dimensions can be controlled through the size of the syringe housing and through the mass of carbon nanotubes added. Their thicknesses are typically much larger than surfactant-cast buckypaper and have been synthesized from 120 μm up to 650 μm; whilst no nomenclature system exists to govern thicknesses for samples to be classified as paper, samples with thicknesses greater than 500 μm are referred to as buckydiscs. The frit compression method allows rapid casting of buckypaper and buckydiscs with recovery of the casting solvent and control over the 2D and 3D geometry.

Aligned multi-walled carbon nanotube (MWCNT) growth has been used in CNT film synthesis through the domino effect. In this process, "forests" of MWCNTs are pushed flat in a single direction, compressing their vertical orientation into the horizontal plane, which results in the formation of high-purity buckypaper with no further purification or treatment required. By comparison, when a buckypaper sample was formed from the 1 ton compression of chemical vapor deposition (CVD) generated MWCNT powder, any application of a solvent led to the immediate swelling of the film till it reverted into particulate matter. It appears that for the CNT powder used, compression alone was insufficient to generate robust buckypaper and highlights that the aligned growth methodology generates *in-situ* tube-tube interactions not found in CVD CNT powder and are preserved through to the domino pushing formation of buckypaper.

Properties

Buckypaper is one tenth the weight yet potentially 500 times stronger than steel when its sheets are stacked to form a composite. It could disperse heat like brass or steel and it could conduct electricity like copper or silicon. Rice University scientist Wade Adams says, "All those things are what a lot of people in nanotechnology have been working toward as sort of Holy Grails."

Applications

Among the possible uses for buckypaper that are being researched:

- Fire protection: covering material with a thin layer of buckypaper significantly improves its fire resistance due to the efficient reflection of heat by the dense, compact layer of carbon nanotubes or carbon fibres.
- If exposed to an electric charge, buckypaper could be used to illuminate computer and television screens. It could be more energy-efficient, lighter, and could allow for a more uniform level of brightness than current cathode ray tube (CRT) and liquid crystal display (LCD) technology.
- Since individual carbon nanotubes are one of the most thermally conductive materials known, buckypaper lends itself to the development

of heat sinks that would allow computers and other electronic equipment to disperse heat more efficiently than is currently possible. This, in turn, could lead to even greater advances in electronic miniaturization.

- Films also could protect electronic circuits and devices within airplanes from electromagnetic interference, which can damage equipment and alter settings. Similarly, such films could allow military aircraft to shield their electromagnetic "signatures", which can be detected via radar.
- Buckypaper could act as a filter membrane to trap microparticles in air or fluid. Because the nanotubes in buckypaper are insoluble and can be functionalized with a variety of functional groups, they can selectively remove compounds or can act as a sensor.
- Produced in high enough quantities and at an economically viable price, buckypaper composites could serve as an effective armor plating.
- Buckypaper can be used to grow biological tissue, such as nerve cells. Buckypaper can be electrified or functionalized to encourage growth of specific types of cells.
- The Poisson's ratio for carbon nanotube buckypaper can be controlled and has exhibited auxetic behaviour, capable of use as artificial muscles.

Carbon Nanocone

Carbon nanocones are conical structures which are made predominantly from carbon and which have at least one dimension of the order one micrometre or smaller. Nanocones have height and base diameter of the same order of magnitude; this distinguishes them from tipped nanowires which are much longer than their diameter. Nanocones occur on the surface of natural graphite. Hollow carbon nanocones can also be produced by decomposing hydrocarbons with a plasma torch. Electron microscopy reveals that the opening angle (apex) of the cones is not arbitrary, but has preferred values of approximately 20°, 40°, and 60°. This observation was explained by a model of the cone wall composed of wrapped graphene sheets, where the geometrical requirement for seamless connection naturally accounted for the semi-discrete character and the absolute values of the cone angle. A related carbon nanoform is the Single-walled carbon nanohorn which typically form aggregates 80-100nm in size.

Free-Standing Hollow Cones

History and Synthesis: Carbon nanocones are produced in an industrial process that decomposes hydrocarbons into carbon and hydrogen with a plasma torch having a plasma temperature above 2000 °C. This method is often referred to as Kvaerner Carbon Black & Hydrogen Process (CBH) and it is relatively "emission-free", i.e., produces rather small amount of air pollutants. At certain, well optimized and patented conditions, the solid carbon output consists of approximately 20% carbon nanocones, 70% flat carbon discs and 10% carbon black.

Plasma-assisted decomposition of hydrocarbons has long been known and applied, for example, for production of carbon fullerenes. Even if not optimized, it yields small amounts of carbon nanocones, which had been directly observed with an electron microscope already in 1994, and their atomic structure was theoretically modelled the same year.

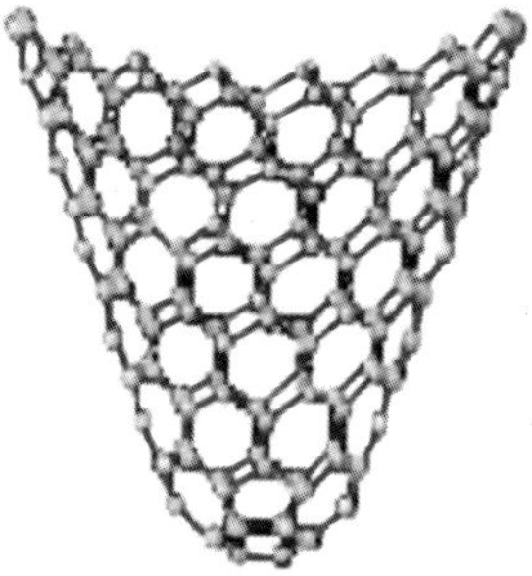

Figure: *Atomic model of a cone with the 38.9° apex angle.*

Modelling

The open carbon cone can be modelled as a wrapped graphene sheet. In order to have strain-free, seamless wrapping, a sector has to be cut out of the sheet. That sector should have an angle of $n \times 60°$, where $n = 1, ..., 5$. Therefore, the resulting cone angle should have only certain, discrete values $\alpha = 2 \arcsin(1 - n/6) = 112.9°, 83.6°, 60.0°, 38.9°$, and $19.2°$ for $n = 1, ..., 5$, respectively. The graphene sheet is composed solely of carbon hexagons which can not form a continuous cone cap. As in the fullerenes, pentagons have to be added to form a curved cone tip, and their number is correspondingly $n = 1, ..., 5$.

Observation

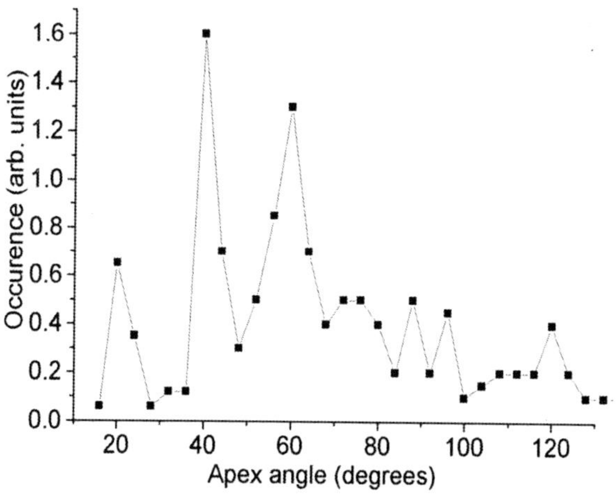

Figure: *Statistical distribution of the apex values measured over 1700 hollow nanocones.*

Electron microscopy observations do confirm the model prediction of discrete cone angles. Two experimental artifacts must be considered though: (i) charging of the poorly conducting carbon samples under electron beam, which blurs the images and (ii) that electron microscopy observations at a fixed sample tilt only yield a two-dimensional projection whereas a 3D shape is required. The first obstacle is overcome by coating the cones with a metal layer of a few nanometres thickness. The second problem is solved through a geometrical shape analysis. Combined with significant statistics on the number of cones, it yields semi-discrete apex angles. Their values deviate from prediction by about 10% due to the limited measurement accuracy and slight variation of the cone thickness along its length.

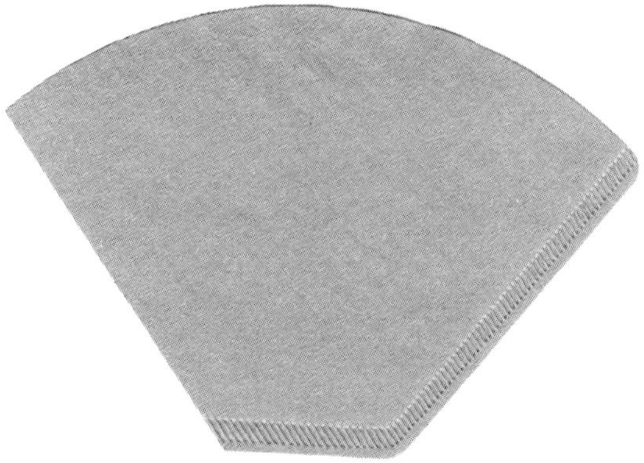

***Figure:** Image of a coffee filter illustrating one of the anomalous structures in the carbon nanocone growth.*

The absolute value of the cone wall thickness varies between 10 and 30 nm, but can be as large as 80 nm for some nanocones. To elucidate the structure of the cone walls, electron diffraction patterns were recorded at different cone orientations. Their analysis suggests that the walls contain 10–30% of ordered material covered with amorphous carbon. High-resolution electron microscopy reveals that the ordered phase consists of nearly parallel layers of graphene. The amorphous fraction can be converted into well-ordered graphite by annealing the cones at temperatures about 2700 °C.

The remarkable feature of the open carbon nanocones produced by the CBH process is their almost ideal shape, with straight walls and circular bases. Non-ideal cones are also observed, but these are exceptions. One such deviation was a "double" cone, which appeared as if a cone started to grow from its tip with a certain apex angle (e.g. 84°), but then abruptly changed the apex angle (e.g. to 39°) in a single point on its surface, thus producing a break in the observed cross-section of the cone. Another anomaly was a cone with the apex extended from a point to a line segment, as in the expanded coffee filter (flat form is shown in the picture).

Other Cones

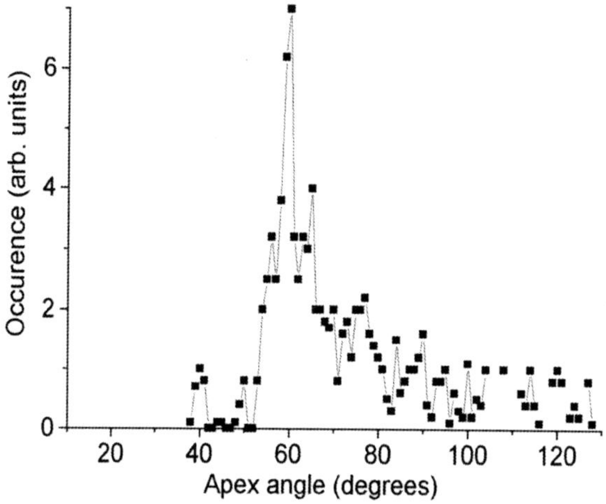

Figure: *Statistical distribution of the apex values measured over 554 cones growing on natural graphite.*

Carbon cones have also been observed, since 1968 or even earlier, on the surface of naturally occurring graphite. Their bases are attached to the graphite and their height varies between less than 1 and 40 micrometres. Their walls are often curved and are less regular than those of the laboratory made nanocones. The distribution of their apex angle also shows a strong feature at 60°, but other expected peaks, at 20° and 40°, are much weaker and the distribution is somewhat broader for large angles. This difference is attributed to the different wall structure of the natural cones. Those walls are relatively irregular and contain numerous line defects (positive-wedge disclinations). This breaks down the angular requirement for a seamless cone and therefore broadens the angular distribution.

Carbon Nanofibres

Carbon nanofibres (CNFs), vapor grown carbon fibres (VGCFs), or vapor grown carbon nanofibres (VGCNFs) are cylindric nanostructures with graphene layers arranged as stacked cones, cups or plates. Carbon nanofibres with graphene layers wrapped into perfect cylinders are called carbon nanotubes.

Carbon is the building block of a myriad of organic and inorganic matter around us. It is a versatile atom capable of joining to other atoms in *sp*, sp^2, and sp^3 hybridized structures giving rise to millions of stable molecules. In its single element form, it has a number of allotropes (polymorphs) like diamond, graphite, and fullerenes with different properties ranging from extremely hard to very soft scope. Carbon can be made to form tubular microstructure called filament or fibre. The unique properties of carbon fibres have expanded the science and technology of composite materials in recent decades. VGCFs (Vapor Grown Carbon Fibre) and their smaller size variant, VGCNFs (Vapor Grown Carbon Nanofibre) are among short carbon fibres that have drawn lots of attention for their potential thermal, electrical, frequency shielding, and mechanical property enhancements. They are being more and more utilised in different material systems like composites thanks to their exceptional properties and low cost.

Synthesis

Catalytic Chemical Vapor Deposition (CCVD) or simply Chemical Vapor Deposition (CVD) with variants like thermal and plasma-assisted is the dominant commercial technique for the fabrication of VGCF and VGCNF. Here, gas-phase molecules are decomposed at high temperatures and carbon is deposited in the presence of a transition metal catalyst on a substrate where subsequent growth of the fibre around the catalyst particles is realised. In general, this process involves separate stages such as gas decomposition, carbon deposition, fibre growth, fibre thickening, graphitization, and purification and results in hollow fibres. The nanofibre diameter depends on the catalyst size. The CVD process for the fabrication of VGCF generally falls into two categories: 1) fixed-catalyst process (batch), and 2) floating-catalyst process (continuous).

In the batch process developed by Tibbetts, a mixture of hydrocarbon/hydrogen/helium was passed over a mullite (crystalline aluminum silicate) with fine iron catalyst particle deposits maintained at 1000 °C. The hydrocarbon used was methane in the concentration of 15% by volume. Fibre growth in several centimetres was achieved in just 10 minutes with a gas residence time of 20 seconds. In general, fibre length can be controlled by the gas residence time in the reactor. Gravity and direction of the gas flow typically affects the direction of the fibre growth.

The continuous or floating-catalyst process was patented earlier by Koyama and Endo and was later modified by Hatano and coworkers. This process typically yields VGCF with sub-micrometre diameters and lengths of a few to 100 μm, which accords with the definition of carbon nanofibres. They utilised organometallic compounds dissolved in a volatile solvent like benzene that would yield a mixture of ultrafine catalyst particles (5-25 nm in diameter) in hydrocarbon gas as the temperature rose to 1100 °C. In the furnace, the fibre growth initiates on the surface of the catalyst particles and continues until catalyst poisoning occurs by impurities in the system. In the fibre growth mechanism described by Baker and coworkers, only the part of catalyst particle exposed to the gas mixture contributes to the fibre growth and the growth stops as soon as the exposed part is covered, i.e. the catalyst is poisoned. The catalyst particle remains buried in the growth tip of the fibre at a final concentration of about a few parts per million. At this stage, fibre thickening takes place.

The most commonly used catalyst is iron, often treated with sulfur, hydrogen sulfide, etc. to lower the melting point and facilitate its penetration into the pores of carbon and hence, to produce more growth sites. Fe/Ni, Ni, Co, Mn, Cu, V, Cr, Mo, Pd, MgO, and Al2O3 are also used as catalyst. Acetylene, ethylene, methane, natural gas, and benzene are the most commonly used carbonaceous gases. Often carbon monoxide (CO) is introduced in the gas flow to increase the carbon yield through reduction of possible iron oxides in the system.

Applications

- Field electron emission sources
- Composite materials
- Scanning probe microscopy tips
- Carrier material for various catalysts in petrochemistry
- In vertically-aligned arrays, a platform for gene delivery.
- For electrode materials
- Oil spill remediation

Colossal Carbon Tube

Colossal carbon tubes (CCTs) are a tubular form of carbon. In contrast to the carbon nanotubes (CNTs), colossal carbon tubes have much larger diameters ranging between 40 and 100 μm. Their walls have a corrugated structure with abundant pores, as in corrugated fibreboard, where the solid membranes have a graphite-like layered structure.

CCTs have technologically attractive properties such as ultralight weight, extremely high strength, excellent ductility and high conductivity - which make them possibly suitable for clothing. They are excellent conductors, are 15 times stronger than the strongest carbon fibre (T1000), have 30 times the tenacity of Kevlar and are 224 times stronger than individual cotton fibres. The tubes exhibit an ultra low density comparable to that of carbon nanofoams.

CCTs have a tensile strength of 7 GPa, and a high specific strength (tensile strength per density), and a breaking length of 6.0×10^3 km. It exceeds the specific strength of the strongest carbon nanotube. This strength is sufficient to support a space elevator if retained in a fabricated macroscale structure.

CCTs conduct electricity and show some of the properties of semiconductors.

Filamentous Carbon

Filamentous carbon is a carbon-containing deposit structure that refers to several allotropes of carbon, including carbon nanotubes, carbon nanofibres, and microcoils. It forms from gaseous carbon compounds. Filamentous carbon structures all contain metal particles. These are either iron, cobalt, or nickel or their alloys. Deposits of it also significantly disrupt synthesis gas methanation. Acetylene is involved in a number of method of the production of filamentous carbon. The structures of filamentous carbon are mesoporous and on the micrometre scale in dimension. Most reactions that form the structures take place at or above 280 °C (536 °F).

Filamentous carbon's applications include cleaning up spills of crude oil and the creation of strong and lightweight composites. Filamentous carbon also has significantly different thermodynamic properties from graphite, another form of carbon. However, filamentous carbon partially consists of graphite sheets.

Properties

Filamentous carbon structures typically range between 10 and 500 nanometres in diameter. They are typically 10,000 nanometres (10 micrometres) long. They have a duplex structure. The outside of the structures is difficult to oxidize, but the core oxidizes more easily. A metal particle is typically located at the growing end of the structure, although it is sometimes found in the middle of it instead. Also, more than one filament can sometimes grow out of one metal particle. Filamentous carbon is either helical, straight, or twisted. It contains graphite layers in a conical shape. They planes of graphite located close to the interface between the carbon and the nickel atom in the filament are parallel to the interface. The filaments can also be hollow.

Filamentous carbon has thermodynamic properties that are different from those of graphite. This is partially due to the fact that the structure of filamentous carbon is more disordered than the structure of graphite. Other theories on the difference in properties include that the deviation is based on the formation of an intermediate phase of carbides. This theory was proposed by De Bokx *et al.* and Manning *et al.*. However, it is unlikely if the central metal atom is nickel because in that case, since carbides decompose at 350 °C (662 °F), and carbides formation was not observed during such a reaction.

A difference between the solubities of metal and filamentous carbon also allows carbon diffusion to occur. When the allotrope engages in a gasification reaction below 600 °C (1,112 °F), the reaction's activation energy is approximately 178 kilojoules per mole.

Filamentous carbon is mesoporous and has unusual textural properties. It also has paramagnetic properties. It also has a high level of mechanical strength.

The nickel particles located in filamentous carbon that is grown in methane and hydrogen gas between 490 °C (914 °F) and 590 °C (1,094 °F) tend to be pear-shaped at the higher end of the temperature range. At higher temperatures, the metal particle becomes deformed. The length of the conic structure of the filaments also increases with temperature. When a copper and silica catalyst is exposed to methane and hydrogen at 927 °C (1,701 °F), hollow, long filamentous carbon structures were formed, and these also contained drops of metal.

Biological Properties

When the enzyme glucoamylase is situated on a ceramic surface coated with filamentous carbon, the enzyme's stability increases drastically.

Occurrence

Filamentous carbon typically forms on metals, including iron, cobalt, and nickel. Hydrogen is also required for filamentous carbon to form. However, thew also form on alloys of these metals. Iron is a better material for forming filamentous carbon on than nickel is. For instance, in the presence of methanol, at a pressure of 7 kilopascals and a temperature of 500 °C (932 °F), filamentous carbon grows on iron, but not nickel. Formation of the material on those metals typically occurs at temperatures between 327 °C (621 °F) and 1,027 °C (1,881 °F). It also forms when chromium is used as a catalyst to decompose acetylene. Filamentous carbon is also one of up to seven allotropes of carbon to form during coke formation on reactor tubes and catalysts. The allotrope has the ability to destroy catalyst support structures, thus blocking reactors. It also forms during stream reforming, along with other varieties of carbon.

Synthesis

Filamentous carbon can also be synthesized by cracking methane. The product is then gasified by hydrogen. In the experiment that discovered this, a nickel particle was used as the metal particle for the filament. The filament precipitates on the nickel particle's "support side".

Filamentous carbon can also form when acetylene decomposes on films of palladium and silicon dioxide. However, filamentous carbon does not form on the palladium and silicon dioxide films if they are preheated with hydrogen at temperatures of 597 °C (1,107 °F). This is because in those conditions, the palladium and silicon dioxide react to form palladium silicide. Iron and silicon dioxide together also act as a catalyst for formation of the structures. Sometimes, iron silicate forms during this reaction.

Another experiment showed that filamentous carbon can form when carbon monoxide is disproprionated over cobaltosic oxide. In the experiment, a heated carburizing gas mixture was sent over powdered cobaltosic oxide. Filamentous carbon is the main deposit that forms from this reaction when it is carried out at 600 °C (1,112 °F).

When chlorobenzene is hydrodechlorinated over nickel and silica, highly ordered structures of filamentous carbon form. When potassium and bromine are present, this reaction can occur at temperatures as low as 280 °C (536 °F). This is because the potassium and bromine aided in restructuring the active sites, thus causing destructive chemisorption of the reactant and also causing the a precipitate of carbon to form. Adding potassium hydroxide to the mixture of nickel and silica in the reaction made little change to the yield of the reaction. However, the addition of potassium bromide significantly increased the yield. Other alkali metal bromides also allow the reaction and the formation of filamentous carbon to occur. Such alkali metal bromides include cesium bromide.

Filamentous carbon can also be synthesized by decomposing chromium carbide at 100 to 200 megapascals and 350 °C (662 °F) to 800 °C (1,470 °F). It has also been formed with a catalyst of cobalt and aluminum phosphate at 2 megapascals and 220 °C (428 °F) to 240 °C (464 °F). The presence of ruthenium in this reaction lessens the yield of filamentous carbon.

Graphene Oxide Paper

Graphene oxide paper or graphite oxide paper is a composite material fabricated from graphite oxide. Micrometre thick films of graphene oxide paper are also named as graphite oxide membranes (in 60-es) or (more recently) graphene oxide membranes. The membranes are typically obtained by slow evaporation of graphene oxide solution or by filtration method.

The material has exceptional stiffness and strength, due to the intrinsic strength of the two-dimensional graphene backbone and to its interwoven layer structure which distributes loads.

The starting material is water-dispersed graphene oxide flakes, which typically contain a single graphene layer. These flakes may be chemically bonded, leading to the development of additional new materials. Like the starting material, graphene oxide paper is an electrical insulator; however, it may be possible to tune this property, making the paper a conductor or semiconductor, without sacrificing its mechanical properties.

Detailed studies of graphite oxide paper by V. Kohlschütter and P. Haenni date back to 1918.

Detailed studies of graphite oxide membranes were performed by P.-H. Boehm (German scientist who invented term "graphene") back in 1960. The paper titled "Graphite Oxide and its membrane properties" reported synthesis of "paper-like foils" with 0.05 mm thickness. The membranes were reported to be not permeable by gases (nitrogen and oxygen) but easily permeable by water vapors and, suggestively, by any other solvents which are able to intercalate graphite oxide. It was also reported that the membranes are not permeable by "substances of lower molecular weight". Permeation of water trough the membrane was attributed to swelling of graphite oxide structure which enables water penetration path between individual graphene oxide layers. The interlayer distance of dried Hummers graphite oxide was reported as 6.35Å but in liquid water it increased to 11.6Å. Remarkably, the paper also cited the inter-layer distance in diluted NaOH as infinity thus reporting dispersion of graphite oxide on single-layered graphene oxide sheets in solution. The study also reported permeation rate of membranes for water 0.1 mg per minute per square cm. The diffusion rate of water was evaluated as 1 cm/hour. H.-P.Boehm's paper also shows that graphite oxide can be used as cation exchange membrane and reports measurements of osmotic pressures, membrane potentials in KCl, HCl, CaC12, MgC12, BaC12 solutions. The membranes were also reported to be permeable by large alkaloid ions as they are able to penetrate between graphene oxide layers.

In 2012 some of the properties of graphite oxide membranes discovered by H.P.Boehm were re-discovered for graphene oxide membranes (essentially the same material with new name): the membranes were reported to be not permeable by helium but permeable by water vapors. This study was later expanded to demonstrate that several salts (for example KCl, MgC12) diffuse trough the graphene oxide membrane if it is immersed in water solution.

Graphene oxide membranes were also actively studied in 60-s for application in water desalination but it never come to practical applications. Retention rates over 90% were reported in this study for NaCl solutions using stabilized graphene oxide membranes in reverse osmosis setup.

8

Molecular Modelling

Molecular modelling encompasses all theoretical methods and computational techniques used to model or mimic the behaviour of molecules. The techniques are used in the fields of computational chemistry, drug design, computational biology and materials science for studying molecular systems ranging from small chemical systems to large biological molecules and material assemblies. The simplest calculations can be performed by hand, but inevitably computers are required to perform molecular modelling of any reasonably sized system. The common feature of molecular modelling techniques is the atomistic level description of the molecular systems. This may include treating atoms as the smallest individual unit (the Molecular mechanics approach), or explicitly modelling electrons of each atom (the quantum chemistry approach).

Molecular Mechanics

Molecular mechanics is one aspect of molecular modelling, as it refers to the use of classical mechanics/Newtonian mechanics to describe the physical basis behind the models. Molecular models typically describe atoms (nucleus and electrons collectively) as point charges with an associated mass. The interactions between neighbouring atoms are described by spring-like interactions (representing chemical bonds) and van der Waals forces. The Lennard-Jones potential is commonly used to describe van der Waals forces. The electrostatic interactions are computed based on Coulomb's law. Atoms are assigned coordinates in Cartesian space or in internal coordinates, and can also be assigned velocities in dynamical simulations. The atomic velocities are related to the temperature of the system, a macroscopic quantity. The collective mathematical expression is known as a potential function and is related to the system internal energy (U), a thermodynamic quantity equal to the sum of potential and kinetic energies. Methods which minimize the potential energy are known as energy minimization techniques (e.g., steepest descent and conjugate gradient), while methods that model the behaviour of the system with propagation of time are known as molecular dynamics.

$$E = E_{\text{bonds}} + E_{\text{angle}} + E_{\text{dihedral}} + E_{\text{non-bonded}}$$

$$E_{\text{non-bonded}} = E_{\text{electrostatic}} + E_{\text{van der Waals}}$$

This function, referred to as a potential function, computes the molecular potential energy as a sum of energy terms that describe the deviation of bond lengths, bond angles and torsion angles away from equilibrium values, plus terms for non-bonded pairs of atoms describing van der Waals and electrostatic interactions. The set of parameters consisting of equilibrium bond lengths, bond angles, partial charge values, force constants and van der Waals parameters are collectively known as a force field.

Different implementations of molecular mechanics use different mathematical expressions and different parameters for the potential function. The common force fields in use today have been developed by using high level quantum calculations and/or fitting to experimental data. The technique known as energy minimization is used to find positions of zero gradient for all atoms, in other words, a local energy minimum.

Lower energy states are more stable and are commonly investigated because of their role in chemical and biological processes. A molecular dynamics simulation, on the other hand, computes the behaviour of a system as a function of time. It involves solving Newton's laws of motion, principally the second law, $F = ma$. Integration of Newton's laws of motion, using different integration algorithms, leads to atomic trajectories in space and time. The force on an atom is defined as the negative gradient of the potential energy function. The energy minimization technique is useful for obtaining a static picture for comparing between states of similar systems, while molecular dynamics provides information about the dynamic processes with the intrinsic inclusion of temperature effects.

Variables

Molecules can be modelled either in vacuum or in the presence of a solvent such as water. Simulations of systems in vacuum are referred to as *gas-phase* simulations, while those that include the presence of solvent molecules are referred to as *explicit solvent* simulations. In another type of simulation, the effect of solvent is estimated using an empirical mathematical expression; these are known as *implicit solvation* simulations.

Applications

Molecular modelling methods are now routinely used to investigate the structure, dynamics, surface properties and thermodynamics of inorganic, biological and polymeric systems. The types of biological activity that have been investigated using molecular modelling include protein folding, enzyme catalysis, protein stability, conformational changes associated with biomolecular function, and molecular recognition of proteins, DNA, and membrane complexes.

Cheminformatics

Cheminformatics (also known as chemoinformatics, chemioinformatics and chemical informatics) is the use of computer and informational techniques applied to a range of problems in the field of chemistry. These *in silico* techniques are used in, for example, pharmaceutical companies in the process of drug discovery. These methods can also be used in chemical and allied industries in various other forms.

Basics

Cheminformatics combines the scientific working fields of chemistry, computer science and information science for example in the areas of topology, chemical graph theory, information retrieval and data mining in the chemical space. Cheminformatics can also be applied to data analysis for various industries like paper and pulp, dyes and such allied industries.

Applications

Storage and Retrieval: The primary application of cheminformatics is in the storage, indexing and search of information relating to compounds. The efficient search of such stored information includes topics that are dealt with in computer science as data mining, information retrieval, information extraction and machine learning. Related research topics include:

- Unstructured data
    - Information retrieval
    - Information extraction
- Structured Data Mining and mining of Structured data
    - Database mining
    - Graph mining
    - Molecule mining
    - Sequence mining
    - Tree mining
- Digital libraries

File Formats

The *in silico* representation of chemical structures uses specialised formats such as the XML-based Chemical Markup Language or SMILES. These representations are often used for storage in large chemical databases. While some formats are suited for visual representations in 2 or 3 dimensions, others are more suited for studying physical interactions, modelling and docking studies.

Virtual Libraries

Chemical data can pertain to real or virtual molecules. Virtual libraries of compounds may be generated in various ways to explore chemical space and hypothesize novel compounds with desired properties.

Virtual libraries of classes of compounds (drugs, natural products, diversity-oriented synthetic products) were recently generated using the FOG (fragment optimized growth) algorithm. This was done by using cheminformatic tools to train transition probabilities of a Markov chain on authentic classes of compounds, and then using the Markov chain to generate novel compounds that were similar to the training database.

Virtual Screening

In contrast to high-throughput screening, virtual screening involves computationally screening *in silico* libraries of compounds, by means of various methods such as docking, to identify members likely to possess desired properties such as biological activity against a given target. In some cases, combinatorial chemistry is used in the development of the library to increase the efficiency in mining the chemical space. More commonly, a diverse library of small molecules or natural products is screened.

Quantitative Structure-Activity Relationship (QSAR)

This is the calculation of quantitative structure-activity relationship and quantitative structure property relationship values, used to predict the activity of compounds from their structures. In this context there is also a strong relationship to Chemometrics. Chemical expert systems are also relevant, since they represent parts of chemical knowledge as an *in silico* representation.

Ninithi (Nanotube Modelling Software)

Ninithi is free and open source modelling software that can be used to visualize and analyze carbon materials used in nanotechnology. Users of ninithi can visualize the 3D molecular geometries of graphene/nano-ribbons, carbon nanotubes (both single wall and multi-wall) and fullerenes. Ninithi also provides features to simulate the electronic band structures of graphene and carbon nanotubes. The software was developed by Lanka Software Foundation, in Sri Lanka and released in 2010 under the GPL licence. Ninithi is written in the Java programming language and available for both Microsoft Windows and Linux platforms.

Generalized equations and algorithms used in ninithi were published in 2010.

Nanoflower

A nanoflower, in chemistry, refers to a compound of certain elements that results in formations which in microscopic view resemble flowers or, in some cases, trees that are called nanobouquets or nanotrees. These formations are nanometres long and thick so they can only be observed using electron microscopy.

Production

Several ways to produce nanoflowers are known:

- A process similar to the making of a carbon nanotube using a hydrocarbon gas.
- Heating gallium (Ga) and then flowing methane (CH_4) over, under specific pressure and heat. This forms flower-shaped silicon carbide (SiC) structures.
- Heating a molybdenum dioxide (MoO_2) thin film on a piece of molybdenum foil surrounded by sulfur vapour.

Nanomeadow

In supercapacitors, energy is stored because the electrodes are coated with a porous material that soaks up ions like a sponge, usually activated carbon. Nanomeadow supercapacitors store ions in manganese oxide (MnO), a material with a much greater capacity for ions than activated carbon.

Scientists at Research Institute of Chemical Defence (Beijing, China) and Peking University created a nanomeadow of microscopic structures, fuzzy flowers of MnO each about 100 nanometres across on a field of messy carbon nanotube grass grown on a tantalum metal foil. Nanomeadows perform 10 times better than MnO alone and can store twice as much charge as the carbon-based electrodes in existing ultracapacitors.

Organic Semiconductor

An organic semiconductor is an organic material with semiconductor properties, that is, with an electrical conductivity between that of insulators and that of metals. Single molecules, oligomers, and organic polymers can be semiconductive. Semiconducting small molecules (aromatic hydrocarbons) include the polycyclic aromatic compounds pentacene, anthracene, and rubrene.

Polymeric organic semiconductors include poly(3-hexylthiophene), poly(p-phenylene vinylene), as well as polyacetylene and its derivatives.

There are two major overlapping classes of organic semiconductors. These are organic charge-transfer complexes and various linear-backbone conductive polymers derived from polyacetylene. Linear backbone organic semiconductors include polyacetylene itself and its derivatives polypyrrole, and polyaniline. At least locally, charge-transfer complexes often exhibit similar conduction mechanisms to inorganic semiconductors. Such mechanisms arise from the presence of hole and electron conduction layers separated by a band gap. Although such classic mechanisms are important locally, as with inorganic amorphous semiconductors, tunnelling, localized states, mobility gaps, and phonon-assisted hopping also significantly contribute to conduction, particularly in polyacetylenes. Like inorganic semiconductors, organic semiconductors can be doped. Organic semiconductors susceptible to doping such as polyaniline (Ormecon) and PEDOT:PSS are also known as *organic metals*.

Processing

There are significant differences between the processing of small molecule organic semiconductors and semiconducting polymers. Thin films of soluble conjugated polymers can be prepared by solution processing methods. On the other hand, small molecules are quite often insoluble and typically require deposition via vacuum sublimation. Both approaches yield amorphous or polycrystalline films with variable degree of disorder. "Wet" coating techniques require polymers to be dissolved in a volatile solvent, filtered and deposited onto a substrate. Common examples of solvent-based coating techniques include drop casting, spin-coating, doctor-blading, inkjet printing and screen printing. Spin-coating is a widely used technique for small area thin film production. It may result in a high material loss. The doctor-blade technique has a minimal material loss and was primarily developed for large area thin film production. Vacuum based thermal deposition of small molecules requires evaporation of molecules from a hot source. The molecules are then transported through vacuum onto a substrate. Condensation of these molecules on the substrate surface results in thin film formation. Wet coating techniques can be applied to small molecules but to a lesser extent depending on material solubility.

Characterization

Organic semiconductors differ from inorganic counterparts in many ways. These include optical, electronic, chemical and structural properties. In order to design and model the organic semiconductors, such optical properties as absorption and photoluminescence need to be characterized. Optical characterization for this class of materials can be done using UV-visible absorption spectrophotometres and photoluminescence spectrometres.

Semiconductor film appearance and morphology can be studied with atomic force microscopy (AFM) and scanning electron microscopy (SEM). Electronic properties such as ionisation potential can be characterized by probing the electronic band structure with ultraviolet photoelectron spectroscopy (UPS).

The charge-carrier transport properties of organic semiconductors are examined by a number of techniques. For example, time-of-flight (TOF) and space charge limited current techniques are used to characterize "bulk" conduction properties of organic films. Organic field effect transistor (OFET) characterization technique is probing "interfacial" properties of semiconductor films and allows to study the charge carrier mobility, transistor threshold voltage and other FET parameters. OFETs development can directly lead to novel device applications such as organic-based flexible circuits, printable radio frequency identification tags (RFID) and active matrix backplanes for displays. Chemical composition and structure of organic semiconductors can be characterized by infrared spectroscopy, secondary ion mass spectrometry (SIMS) and X-ray photoelectron spectroscopy (XPS).

Charge Transport in Disordered Organic Semiconductors

Charge transport in organic semiconductors is dependent on π-bonding orbitals and quantum mechanical wave-function overlap. In disordered organic semiconductors, there is limited π-bonding overlapping between molecules and conduction of charge carriers (electrons or holes) is described by quantum mechanical tunnelling. Charge transport depends on the ability of the charge carriers to pass from one molecule to another. Because of the quantum mechanical tunnelling nature of the charge transport, and its subsequent dependence on a probability function, this transport process is commonly referred to as hopping transport. Hopping of charge carriers from molecule to molecule depends upon the energy gap between HOMO and LUMO levels. Carrier mobility is reliant upon the abundance of similar energy levels for the electrons or holes to move to and hence will experience regions of faster and slower hopping. This can be affected by both the temperature and the electric field across the system.

A theoretical study has shown that in a low electric field the conductivity of organic semiconductor is proportional to $T^{-1/4}$ and in a high electric field is proportional to $e^{-(E/\alpha T)}$, where α is a constant of the material. Another study shows that the AC conductivity of the organic semiconductor pentacene is frequency-dependent and provided evidence that this behaviour is due to its polycrystalline structure and hopping conduction.

Selective Chemistry of Single-Walled Nanotubes

Selective chemistry of single-walled nanotubes is a field in Carbon nanotube chemistry devoted specifically to the study of functionalization of single-walled carbon nanotubes.

Structure and Reactivity

Reactivity of fullerene molecules with respect to addition chemistries is strongly dependent on the curvature of the carbon framework. Their outer surface (exohedral) reactivity increases with increase in curvature. In comparison with fullerene molecules single-walled nanotubes (SWNTs) are moderately curved. Consequently nanotubes are expected to be less reactive than most fullerene molecules due to their smaller curvature, but more reactive than a graphene sheet due to pyramidalization and misalignment of pi-orbitals. The strain of a carbon framework is also reflected in the pyramidalization angle (θ_p) of the carbon constituents. Trigonal carbon atoms (sp^2 hybridized) prefer a planar orientation with θ_p=0° (i.e. graphene) and fullerene molecules have θ_p= 11.6°. The (5,5) SWNT has θ_p~6° for the sidewall. Values for other (n,n) nanotubes show a trend of increasing θ_p (sidewall) with decrease in n. Therefore generally the chemical reactivity of SWNT increases with decrease in diameter (or n, diameter increases with n). Apart from the curvature SWNT reactivity is also highly sensitive to chiral wrapping (n,m) which determine its electronic structure. Nanotubes with $n - m = 3i$ (i is an integer) are all metals and rest are all semiconducting (SC).

Sidewall Functionalization

Carbon nanotubes are metallic or semiconducting, based upon delocalized electrons occupying a 1-D density of states. However, any covalent bond on SWNT sidewall causes localization of these electrons. In the vicinity of localized electrons, the SWNT can no longer be described using a band model that assumes delocalized electrons moving in a periodic potential.

Two important addition reactions of SWNT sidewall are: (1) Fluorination, and (2) Aryl diazonium salt addition. These functional groups on SWNT improve solubility and processibility. Moreover, these reactions allow for combining unique properties of SWNTs with those of other compounds. Above all, the selective diazonium chemistry can be used to separate the semiconducting and metallic nanotubes.

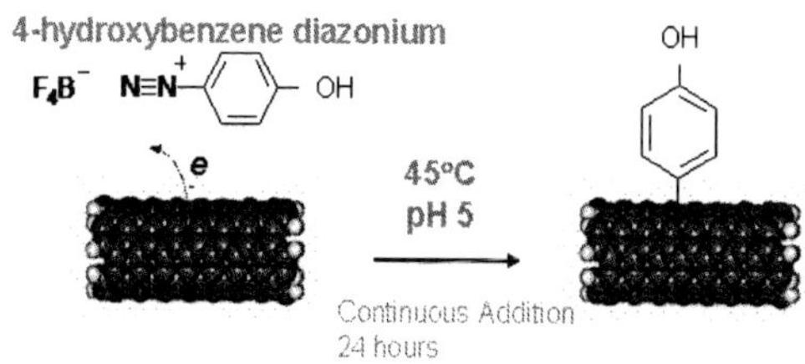

Selective Diazonium Reaction

pH is reduced to prevent electrostatic repulsion.

Temperature is increased to increase the reaction rate.

Diazonium salt is added at a very small rate

Fluorination

The first extensive SWNT sidewall reaction was fluorination in 1998 by Mickleson et al. These fluorine moieties can be removed from the nanotube by treatment in hydrazine and the spectroscopic properties of the SWNT can be restored completely.

Diazonium Chemistry

One of the most important SWNT sidewall reaction is that with diazonium reagent which if done under controlled conditions can be used to do selective covalent chemistry.

Water-soluble diazonium salts react with carbon nanotubes via charge transfer in which they extract electrons from SWNT and form a stable covalent aryl bond. This covalent aryl bond forms with extremely high affinity for electrons with energies near the Fermi level, E_f of the nanotube. Metallic SWNT have a greater electron density near E_f resulting in their higher reactivity over semiconducting nanotubes. The reactant forms a charge-transfer complex at the nanotube surface, where electron donation from the latter stabilizes the transition state and accelerates the forward rate. Once the bond symmetry of the nanotube is disrupted by the formation of this defect, adjacent carbons increase in reactivity and initial selectivity for metallic SWNT is amplified. Under carefully controlled conditions this behaviour can be exploited to obtain highly selective functionalization of metallic nanotubes to the near exclusion of the semiconductors.

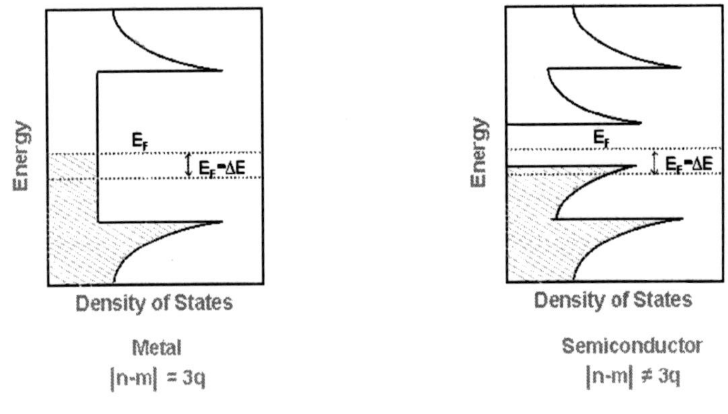

Density of States Control the Selectivity of Covalent Bond Formation

- Selectivity is dictated by the availability of electrons near the Fermi level to stabilize a charge-transfer transition state preceding bond formation. Covalent bond forms with extremely high affinity for electrons with energies within ΔE, near the Fermi level, E_f of the nanotube.

Selective Reaction Conditions

Primary condition is addition of reactant molecules at a very small rate to SWNT solution for a sufficient long time. This ensures reaction with only metallic SWNTs and with no semiconducting SWNTs as all the reactant molecules are taken up by the metallic SWNTs. Long time injection ensures that all metallic tubes are reacted. For example one highly selective condition is: addition of 500 μL of 4-hydroxybenzene diazonium tetrafluoroborate solution in water (0.245 mM) at an injection rate of 20.83 μL/h into 5 mL of SWNT solution (1 wt % sodium dodecyl sulfate (SDS)) over 24 hrs. However, if the entire diazonium solution is added all instantaneously then semiconducting SWNTs will also react due to presence of excess reactant.

Spectroscopy and Functionalization

SWNTs have unique optical and spectroscopic properties largely due to one-dimensional confinement of electronic and phonon states, resulting in so-called van Hove singularities in the nanotube density of states (DOS).

Probing Selective Chemistry via Optical Absorption

Optical absorption monitors the valence (v) to conduction (c) electronic transitions denoted E_{nn} where n is the band index. The E_{11} transitions for the metallic nanotubes occur from ~440 to 645 nm. The E_{11} and E_{22} transitions for the semiconducting nanotubes are found from 830 to 1600 nm and 600 to 800 nm, respectively. These separated absorption features allow for the monitoring of valence electrons in each distinct nanotube. Reaction at the surface result in localization of valence electrons makes them no longer free to participate in photoabsorption which results in decay of the spectrum features.

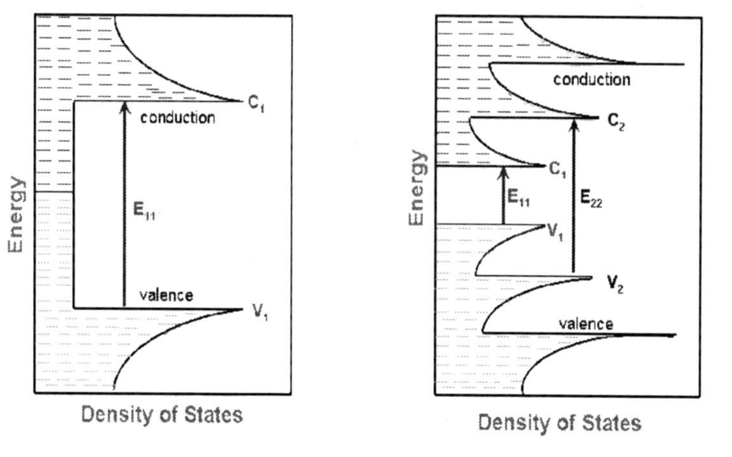

Metallic SWNT

$v_1 \rightarrow c_1$ corresponds to the "first van Hove" optical transition

Semiconducting SWNT

$v_2 \rightarrow c_2$ corresponds to the "second van Hove" optical transition

$v_1 \rightarrow c_1$ corresponds to the "first van Hove" optical transition

Density of Electronic States (DOS) for a Single SWNT Structure

Selective diazonium chemistry abruptly decreases the peak intensities that represent the first Van Hove transition of metallic species (E_{11}, metal), while the peak intensities representing the second (E_{22}, semiconducting) and first (E_{11}, semiconducting) Van Hove transition of the semiconducting species show little or no change. A relative decrease in metallic SWNT absorption features over semiconducting features represents a highly preferential functionalization of the metallic nanotubes.

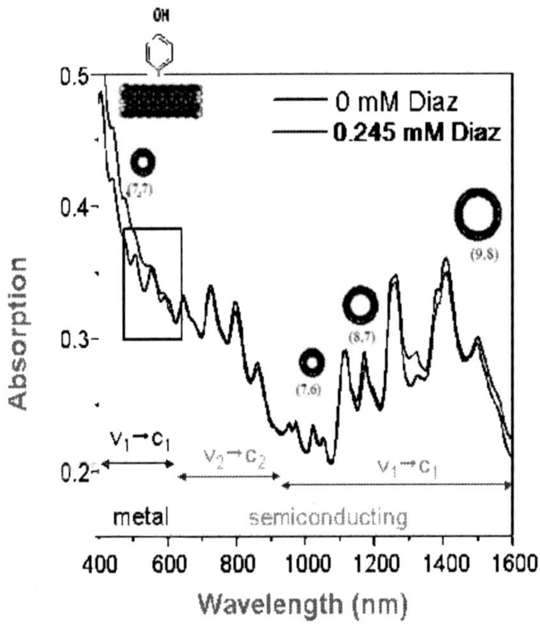

Ultraviolet- visible–near-Infrared (UV-vis-nIR) Absorption Spectrum Allows Probing of Valence Electrons

Absorption spectra of aqueous suspended nanotubes after successive additions of 4-hydroxybenzenediazonium tetrafluoroborate under controlled conditions. Transition for metallic SWNT decay almost completely due to covalent reaction. Transitions for SC nanotubes remain almost unchanges.

Raman Spectroscopy

Raman spectroscopy is a powerful technique with wide ranging applications in carbon nanotube studies. Some important Raman features are radial breathing mode (RBM), tangential mode (G-band), and disorder-related mode.

RBM features correspond to the coherent vibration of the C atoms in the radial direction of the nanotube. These features are unique in carbon nanotubes and occur with the frequencies ω_{RBM} between 120 and 350 cm^{-1} for SWNT in the diameter range (0.7 nm-2 nm). They can be used to probe the SWNT diameter, electronic structure through their frequency and intensity (I_{RBM}) respectively and hence perform an (n,m) assignment to their peaks. The addition of the moiety to the sidewall of the nanotube disrupts the oscillator strength that gives rise to RBM feature and hence causes decay of these features. These features are distinct for species of a particular nanotube (n,m) and hence enables to probe which SWNTs are functionalized and to what extent.

Two main components of the tangential mode include G^+ at 1590 cm^{-1} and G^- at 1570 cm^{-1}. G^+ feature is associated with carbon atom vibrations along the nanotube axis. The G^- feature is associated with vibrations of carbon atoms along the circumferential direction. The G-band frequency can be used (1) to distinguish between metallic and semiconducting SWNTs, and (2) to probe charge transfer arising from doping a SWNT. Frequency of G^+ is sensitive to charge transfer. It upshifts for acceptors and downshifts for donors. Lineshape of G^- is highly sensitive to the whether SWNT is metallic (Breit-Wigner-Fano lineshape) or semiconducting (Lorentzian lineshape).

The disorder-related mode (D peak) is a phonon mode at 1300 cm^{-1} and involves the resonantly enhanced scattering of an electron via phonon emission by a defect that breaks the basic symmetry of the graphene plane. This mode corresponds to the conversion of a sp^2-hybridized carbon to a sp^3-hybridized on the surface. Intensity of D peak measures covalent bond made with the nanotube surface. This feature does not increase as a result of surfactant or hydronium ion adsorption on the nanotube surface.

Selective Reaction and Raman Features

Selective functionalization increases the intensity of the D peak due to formation of aryl-nanotube bond and decreases the tangential mode due loss of electronic resonance. These two effects are generally summed together as increase in their peaks ratio (D/G). RBM peaks of metallic nanotubes decay and the peaks corresponding to those of semiconducting nanotubes remain almost unchanged.

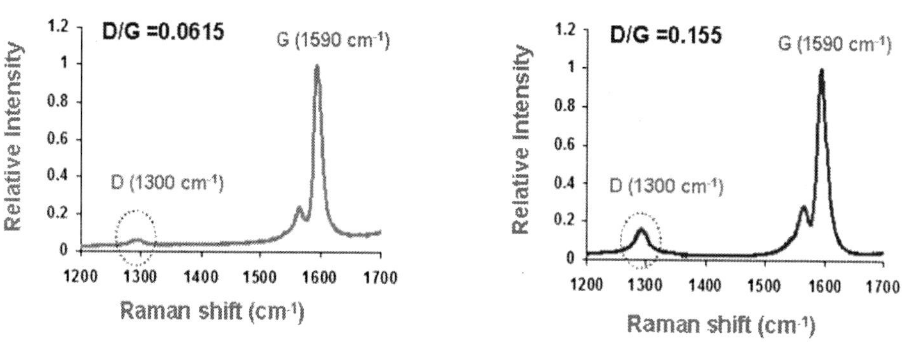

Raman Spectrum and Covalent Reaction

Disorder mode grows with increase in functionalization.

(Raman spectra taken at 785-nm excitation)

Reaction Mechanism

Diazonium reagent and SWNT reaction has a two step mechanism. First, the diazonium reagent adsorbs noncovalently to an empty site on the nanotube surface, forming a charge-transfer complex. This is a fast, selective noncovalent adsorption and diazonium group in this complex partially dopes the nanotube, diminishing the tangential mode in the Raman spectrum. Desorption of A from nanotube is negligible ($k_{-1} \sim 0$). In second step complex B then decomposes to form a covalent bond with the nanotube surface. This is a slower step that need not be selective and is represented by the restoration of the G peak and increase in the D band.

Reversibility of Diazonium Chemistry

Nanotubes reacted with the diazonium reagent can be converted back into pristine nanotubes when thermally treated at 300 °C in an atmosphere of inert gas. This cleaves the aryl hydroxyl moieties from the nanotube sidewall and restores the spectroscopic feature (Raman and absorption spectra) of pristine nanotube.

Chemical Separation of Metallic and Semiconducting SWNTs

Metallic and semiconducting carbon nanotubes generally coexist in as-grown materials. To get only semiconducting or only metallic nanotubes selective functionalization of metallic SWNTs via 4-hydroxybenzene diazonium can be used. Separation can be done in solution by deprotonation of the p-hydroxybenzene group on the reacted nanotubes (metallic) in alkaline solution followed by electrophoretic separation of these charged species from the neutral species (semiconducting nanotubes). This followed by annealing would give separated pristine semiconducting and metallic SWNT.

Metamaterials

Metamaterials are artificial materials engineered to have properties that may not be found in nature. They are assemblies of multiple individual elements fashioned from conventional microscopic materials such as metals or plastics, but the materials are usually arranged in repeating patterns. Metamaterials gain their properties not from their composition, but from their exactingly-designed structures. Their precise shape, geometry, size, orientation and arrangement can affect the waves of light (electromagnetic radiation) or sound in an unconventional manner, creating material properties which are unachievable with conventional materials. These metamaterials achieve desired effects by incorporating structural elements of sub-wavelength sizes, i.e. features that are actually smaller than the wavelength of the waves they affect.

The primary research in metamaterials investigates materials that are able to reverse the refractive index. These materials, known as negative index metamaterials, additionally permit the creation of superlenses that can greatly increase optical resolution beyond the capability of conventional lenses, becoming a solution to the centuries old problem of diffraction-limited systems. In other work, a form of 'invisibility' has been demonstrated at least over a narrow wave band with gradient-index materials. Although the first metamaterials were electromagnetic, acoustic and seismic metamaterials are also areas of active research.

Potential applications of metamaterials are diverse and include remote aerospace applications, sensor detection and infrastructure monitoring, smart solar power management, public safety, radomes, high-frequency battlefield communication and lenses for high-gain antennas, improving ultrasonic sensors, and even shielding structures from earthquakes. The research in metamaterials is interdisciplinary and involves such fields as electrical engineering, electromagnetics, solid state physics, microwave and antennae engineering, optoelectronics, classic optics, material sciences, semiconductor engineering, nanoscience and others.

Electromagnetic Metamaterials

An electromagnetic metamaterial affects electromagnetic waves by having structural features smaller than the wavelength of the respective electromagnetic wave. To behave as a homogeneous material accurately described by an effective refractive index, its features must be much smaller than the wavelength. For microwave radiation, the cells need to be on the order of several millimetres. Microwave frequency metamaterials are usually constructed as arrays of electrically conductive elements (such as loops of wire) that have suitable inductive and capacitive characteristics. These are known as split-ring resonators.

Photonic metamaterials, at the scale of nanometres, are being studied in order to manipulate light at optical frequencies. To date, subwavelength structures have shown only a few questionable results at visible wavelengths. Photonic crystals and frequency-selective surfaces such as diffraction gratings, dielectric mirrors, and optical coatings exhibit similarities to subwavelength structured metamaterials. However, these are usually considered distinct from subwavelength structures, as their features are structured for the wavelength at which they function, and thus cannot be approximated as a homogeneous material. However, material structures such as photonic crystals are effective in the visible light spectrum. The middle of the visible spectrum has a wavelength of approximately 560 nm (for sunlight), the photonic crystal structures are generally half this size or smaller, that is <280 nm.

Plasmonic metamaterials utilise surface plasmons, which are packets of electrical charges that collectively oscillate at the surfaces of metals at optical frequencies.

Frequency selective surfaces (FSS) can exhibit subwavelength characteristics and are known variously as Artificial magnetic conductors (AMC) or High Impedance Surfaces (HIS). FSS display inductive and capacitive characteristics that are directly related to their subwavelength structure.

Negative Refractive Index in Metamaterials

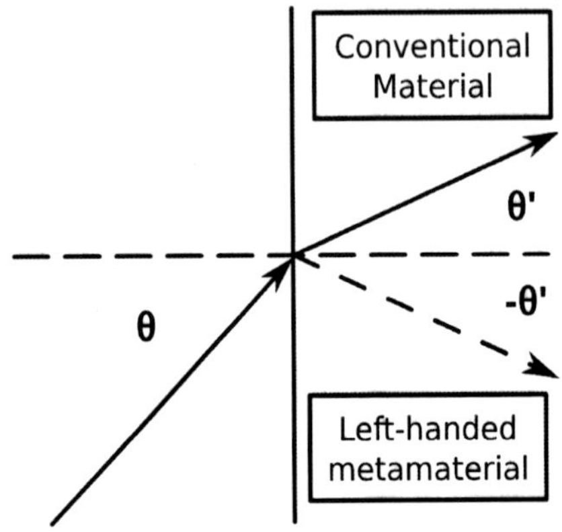

Figure: *A comparison of refraction in a left-handed metamaterial to that in a normal material*

Almost all materials encountered in optics, such as glass or water, have positive values for both permittivity ε and permeability μ. However, metals such as silver and gold have negative permittivity at shorter wavelengths. A material such as a surface plasmon that has either (but not both) ε or μ negative is often opaque to electromagnetic radiation. However, anisotropic materials with only negative permittivity can produce negative refraction due to chirality.

Although the optical properties of a transparent material are fully specified by the parameters εr and μr, refractive index n is often used in practice, which can be determined from $n = \pm\sqrt{\epsilon_r \mu_r}$. All known non-metamaterial transparent materials possess positive εr and μr. By convention the positive square root is used for n.

However, some engineered metamaterials have εr < 0 and μr < 0. Because the product εrμr is positive, n is real. Under such circumstances, it is necessary to take the negative square root for n.

The foregoing considerations are simplistic for actual materials, which must have complex-valued εr and μr. The real parts of both εr and μr do not have to be negative for a passive material to display negative refraction. Metamaterials with negative n have numerous interesting properties:

- Snell's law ($n_1 \sin\theta_1 = n_2 \sin\theta_2$), but as n_2 is negative, the rays will be refracted on the *same* side of the normal on entering the material.
- Cherenkov radiation points the other way.

- The time-averaged Poynting vector is antiparallel to phase velocity. However, for waves (energy) to propagate, a $-\mu$ must be paired with a $-\varepsilon$ in order to satisfy the wave number dependence on the material parameters $k = \omega\sqrt{\mu\epsilon}$.

For plane waves propagating in electromagnetic metamaterials, the electric field, magnetic field and wave vector follow a left-hand rule. This is a reversal of direction when compared to the behaviour of conventional optical materials. Negative refractive index is an important characteristic in metamaterial design and fabrication. As reverse-refraction media, these occur when both permittivity ε and permeability μ are negative. Furthermore, this condition occurs mathematically from the vector triplet E, H and k. In ordinary materials – solid, liquid, or gas; transparent or opaque; conductor or insulator – the conventional refractive index dominates. This means that permittivity and permeability are both positive resulting in an ordinary index of refraction. However, metamaterials have the capability to exhibit a state where both permittivity and permeability are negative, resulting in an extraordinary, index of negative refraction.

Classification

Electromagnetic metamaterials divide into different classes, as follows:

Negative Index

In *negative index metamaterials* (NIM), both permittivity and permeability are negative resulting in a negative index of refraction. These are also known as Double Negative Metamaterials or double negative materials (DNG). Other terms for NIMs include "left-handed media", "media with a negative refractive index", and "backward-wave media". In optical materials, if both permittivity ε and permeability μ are positive, wave propagation travels in the *forward* direction. If both ε and μ are negative, a *backward* wave is produced. If ε and μ have different polarities, waves do not propagate. Mathematically, quadrant II and quadrant IV have coordinates (0,0) in a coordinate plane where ε is the horizontal axis, and μ is the vertical axis. To date, materials exhibiting a negative index of refraction have only been demonstrated as artificially constructed materials.

Single Negative

Single negative (SNG) metamaterials have either negative relative permittivity (εr) or negative relative permeability (μr), but not both. They act as metamaterials in combination with a different, complementary SNG, jointly acting as a DNG.

Epsilon negative media (ENG) display a negative εr while μr is positive. Many plasmas exhibit this characteristic. For example noble metals such as gold or silver will exhibit this characteristic in the infrared and visible spectrums. Mu-negative media (MNG) display a positive εr while μr is negative. *Gyrotropic* or *gyromagnetic* materials exhibit this characteristic. A *gyrotropic* material is one that has been altered by the presence of a quasistatic magnetic field, enabling a *magneto-optic effect*. A *magneto-optic effect* is a phenomenon in which an electromagnetic wave propagates through such a medium. In such a material, left- and right-rotating elliptical polarizations can propagate at different speeds. When light is transmitted through a layer of magneto-optic material, the result is called the Faraday effect: the polarization plane can be rotated, forming a Faraday rotator. The results of such a reflection are known as the magneto-optic Kerr effect (not to be confused with the nonlinear Kerr effect). Two gyrotropic materials with reversed rotation directions of the two principal polarizations are called optical isomers. Joining a slab of ENG material and slab of MNG material resulted in properties such as resonances, anomalous tunneling, transparency, and zero reflection. Like negative index materials, SNGs are innately dispersive, so their εr, μr and refraction index n, alter with changes in frequency.

Electromagnetic Bandgap

Electromagnetic bandgap metamaterials (EBM) control light propagation. This is accomplished either with photonic crystals (PC) or left-handed materials (LHM). PCs can prohibit light propagation altogether. Both classes can allow light to propagate in specific, designed directions and both can be designed with bandgaps at desired frequencies. The period size of EBGs is an appreciable amount of the wavelength, creating constructive and destructive interference. PC are distinguished from sub-wavelength structures, such as tunable metamaterials, because the PC derives its properties from its bandgap characteristics. PCs are sized to match the wavelength of light, versus other metamaterials that operate as a sub-wavelength structure. Furthermore, photonic crystals function by diffracting light. In contrast, a permittivity and permeability define each metamaterial, which is derived from its sub-wavelength structure and does not use diffraction. PCs have periodic inclusions that inhibit wave propagation due to the inclusions' destructive interference from scattering. The photonic bandgap property of PCs makes them the electromagnetic analog of electronic semi-conductor crystals. EBGs have the goal of creating high quality, low loss, periodic, dielectric structures. An EBG affects photons in the same way semiconductor materials affect electrons. PCs are the perfect bandgap material, because they allow no light propagation. Each unit of the prescribed periodic structure acts like one atom, albeit of a much larger size.

EBGs are designed to prevent the propagation of an allocated bandwidth of frequencies, for certain arrival angles and polarizations. Various geometries and structures have been proposed to fabricate EBG's special properties. In practice it is impossible to build a flawless EBG device. EBGs have been manufactured for frequencies ranging from a few gigahertz (GHz) up to a few terahertz (THz), radio, microwave and mid-infrared frequency regions. EBG application developments include a transmission line, woodpiles made of square dielectric bars, and several different types of low gain antennas.

Double Positive Medium

Double positive mediums (DPS) do occur in nature, such as naturally occurring dielectrics. Permittivity and magnetic permeability are both positive and wave propagation is in the forward direction. Artificial materials have been fabricated which combine DPS, ENG and MNG properties.

Bi-Isotropic and Bianisotropic

Categorizing metamaterials into double or single negative, or double positive, normally assumes that the metamaterial has independent electric and magnetic responses described by ε and μ. However in many cases, the electric field causes magnetic polarization, while the magnetic field induces electrical polarization, known as magnetoelectric coupling. Such media are denoted as bi-isotropic. Media that exhibit magnetoelectric coupling and that are also anisotropic (which is the case for many metamaterial structures), are referred to as bi-anisotropic. Four material parameters are intrinsic to magnetoelectric coupling of bi-isotropic media. They are the electric (E) and magnetic (H) field strengths, and electric (D) and magnetic (B) flux densities. These four material parameters are ε, μ, κ and χ or permittivity, permeability, strength of chirality, and the Tellegen parameter respectively. In this type of media, the material parameters do not vary with changes along a rotated coordinate system of measurements. In this sense they are invariant or scalar.

The intrinsic magnetoelectric parameters, κ and χ, affect the phase of the wave. The effect of the chirality parameter is to split the refractive index. In isotropic media this results in wave propagation only if ε and μ have the same sign. In bi-isotropic media with χ assumed to be zero, and κ a non-zero value, different results appear. Both a backward wave and a forward wave can occur. Alternatively, two forward waves or two backward waves can occur, depending on the strength of the chirality parameter.

Chiral

Chiral metamaterials are constructed from chiral in which the effective parameter k is non-zero. This is a potential source of confusion as the metamaterial literature includes two conflicting uses of the terms *left-* and *right-handed*. The first refers to one of the two circularly polarized waves that are the propagating modes in chiral media. The second relates to the triplet of electric field, magnetic field and Poynting vector that arise in negative refractive index media, which in most cases are not chiral. Wave propagation properties in chiral metamaterials demonstrate that negative refraction can be realised in metamaterials with a strong chirality and positive ε and μ. This is because the refractive index has distinct values for left and right, given by

$$n = \sqrt{\epsilon_r \mu_r} \pm \kappa$$

It can be seen that a negative index will occur for one polarization if $\kappa > \sqrt{\varepsilon r \mu r}$. In this case, it is not necessary that either or both εr and μr be negative for backward wave propagation.

FSS Based

Frequency selective surface-based metamaterials block signals in one waveband and pass those at another waveband. They have become an alternative to fixed frequency metamaterials. They allow for optional changes of frequencies in a single medium, rather than the restrictive limitations of a fixed frequency response.

Acoustic

Acoustic metamaterials control, direct and manipulate sound in the form of sonic, infrasonic, or ultrasonic waves in gases, liquids and solids. As with electromagnetic waves, sonic waves can exhibit negative refraction.

Control of sound waves is mostly accomplished through the bulk modulus β, mass density ρ, and chirality. The bulk modulus and density are analogs of permittivity and permeability in electromagnetic metamaterials. Related to this is the mechanics of sound wave propagation in a lattice structure. Also materials have mass and intrinsic degrees of stiffness. Together, these form a resonant system and the mechanical (sonic) resonance may be excited by appropriate sonic frequencies (for example audible pulses).

Basic Principles

Since the acoustic metamaterials are one of the branch of the metamaterials, the basic principle of the acoustic metamaterials is similar to the principle of metamaterials. These metamaterials usually gain their properties from structure rather than composition, using the inclusion of small inhomogeneities to enact effective macroscopic behaviour. Similar to metamaterials research, investigating materials with Negative index metamaterials, the negative index acoustic metamaterials became the primary research. Negative refractive index of acoustic materials can be achieved by changing the bulk modulus and mass density.

Bulk Modulus and Mass Density

Below, the bulk modulus β of a substance reflects the substance's resistance to uniform compression. It is defined in relation to the pressure increase needed to cause a given relative decrease in volume.

The mass density (or just "density") of a material is defined as mass per unit volume and is expressed in grams per cubic centimetre (g/cm^3). In all three classic states of matter — gas, liquid, or solid — the density varies with a change in temperature or pressure, and gases are the most susceptible to those changes. The spectrum of densities is wide ranging: from 10 g/cm^3 for neutron stars, 1.00 g/cm^3 for water to 1.2×10"^{3} g/cm^3 for air. Also relevant here are area density which is mass over a (two-dimensional) area, linear density - mass over a one-dimensional line, and relative density, which is a density divided by the density of a reference material, such as water.

For acoustic materials and acoustic metamaterials, both bulk modulus and density are component parameters, which define their refractive index.

Analogues

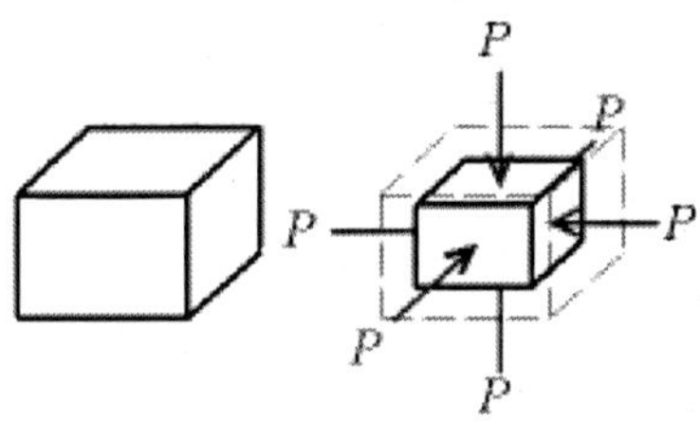

Figure: *Bulk modulus - illustration of uniform compression*

Scientific research revealed that acoustic metamaterials have analogues to electromagnetic metamaterials when exhibiting the following characteristics:

In certain frequency bands, the *effective mass density* and *bulk modulus* may become negative. This results in a *negative refractive index. Flat slab focusing*, which can result in *super resolution*, is similar to electromagnetic metamaterials. The double negative parameters are a result of low-frequency resonances. In combination with a well-defined polarization during wave propagation; $k = |n|\omega$, is an equation for refractive index as sound waves interact with acoustic metamaterials (below):

$$n^2 = \frac{\rho}{\beta}$$

The inherent parameters of the medium are the mass density ρ, bulk modulus β, and chirality *k*. Chirality, or handedness, determines the polarity of wave propagation (wave vector). Hence within the last equation, Veselago-type solutions ($n^2 = u^*\varepsilon$) are possible for wave propagation as the negative or positive state of ρ and β determine the forward or backward wave propagation.

In negative refractive, electromagnetic metamaterials, negative permittivity can be found in natural materials. However, negative permeability has to be intentionally created in the artificial transmission medium. Obtaining a negative refractive index with acoustic materials is different. Neither negative ρ nor negative β are found in naturally occurring materials; they are derived from the resonant frequencies of an artificially fabricated transmission medium (metamaterial), and such negative values are an anomalous response. Negative ρ or β means that at certain frequencies the medium expands when experiencing compression (negative modulus), and accelerates to the left when being pushed to the right (negative density).

Electromagnetic Field vs Acoustic Field

The electromagnetic spectrum extends from below frequencies used for modern radio to gamma radiation at the short-wavelength end, covering wavelengths from thousands of kilometres down to a fraction of the size of an atom. That would be wavelengths from 10^3 to 10^{-15} kilometres. The long wavelength limit is the size of the universe itself, while it is thought that the short wavelength limit is in the vicinity of the Planck length, although in principle the spectrum is infinite and continuous.

Infrasonic frequencies range from 20 Hz down to 0.001 Hz. Audible frequencies are 20 Hz to 20 kHz. Ultrasonic range is above 20 kHz. Sound requires a medium. Electromagnetics radiation (EM waves) can travel in a vacuum.

Mechanics of Lattice Waves

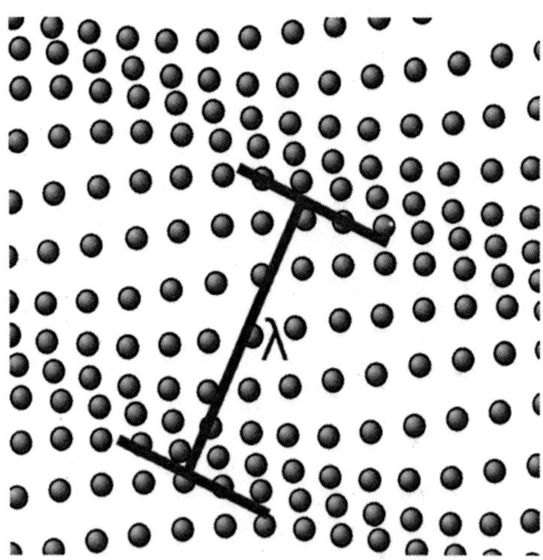

An imaginary demonstration: A hypothetical rigid lattice structure (solid) is composed of 10^{23} atoms. However, in a real solid these particles could just as easily be ions. In a rigid lattice structure, atoms exert pressure, or a force, on each other in order to maintain equilibrium. Atomic forces maintain rigid lattice structure. Most of them, such as the covalent or ionic bonds, are of electric nature. The magnetic force, and the force of gravity are negligible. Because of bonding between atoms, the displacement of one or more atoms from their equilibrium positions will give rise to a set of vibration waves propagating through the lattice. One such wave is shown in the figure to the right. The amplitude of the wave is given by the displacements of the atoms from their equilibrium positions. The wavelength λ is marked.

There is a *minimum possible* wavelength, given by the equilibrium separation *a* between atoms. Any wavelength shorter than this can be mapped onto a wavelength longer than *a*, due to effects similar to that in aliasing.

Analysis and Experiments

The current research on acoustic metamaterials is based not only on prior experience with electromagnetic metamaterials. The key physics in acoustics are *sound*, *ultrasound* and *infrasound*, which are mechanical waves in gases, liquids, and solids. One objective of the inquiry into the properties of acoustic metamaterials is applications in seismic wave reflection and in vibration control technologies related to earthquakes.

Sonic Crystals

In the year 2000 the research of Liu *et al.* paved the way to acoustic metamaterials through sonic crystals. The latter exhibit spectral gaps two orders of magnitude smaller than the wavelength of sound. The spectral gaps prevent the transmission of waves at prescribed frequencies. The frequency can be tuned to desired parameters by varying the size and geometry of the metamaterial.

The fabricated material consisted of a high-density solid lead ball as the core, one centimetre in size, which was coated with a 2.5-mm layer of rubber silicone. These were arranged in a crystal lattice structure of an 8 × 8 × 8 cube. The balls were cemented into the cubic structure with an epoxy. Transmission was measured as a function of frequency from 250 to 1600 Hz for effectively a four-layer sonic crystal. A two-centimetre slab absorbed sound that normally would require a much thicker material, at 400 Hz. A drop in amplitude was observed at 400 and 1100 Hz.

The amplitudes of the sound waves entering the surface were compared with the sound waves at the centre of the metamaterial structure. The oscillations of the coated spheres absorbed sonic energy, which created the frequency gap; the sound energy is absorbed exponentially as the thickness of the material is increased. The key result here is a negative elastic constant created from resonant frequencies of the material. Its projected applications, with a future expanded frequency range in elastic wave systems, are seismic wave reflection and ultrasonics.

Split-Ring Resonators for Acoustic Metamaterials

In 2004 split-ring resonators (SRR) became the object of acoustic metamaterial research. Prior research with SRRs fabricated as negative index *electromagnetic* metamaterials was referenced as the progenitor of further research in acoustic metamaterials. An analysis of the frequency band gap characteristics, derived from the inherent limiting properties of artificially created SRRs, paralleled an analysis of sonic crystals. The band gap properties of SRRs were related to sonic crystal band gap properties. Inherent in this inquiry is a description of mechanical properties and problems of continuum mechanics for sonic crystals, as a macroscopically homogeneous substance.

The correlation in bandgap capabilities includes locally resonant elements and elastic moduli which operate in a certain frequency range. Elements which interact and resonate in their respective localized area are embedded throughout the material. In acoustic metamaterials, locally resonant elements would be the interaction of a single 1-cm rubber sphere with the surrounding liquid. The values of the stop band and band gap frequencies can be controlled by choosing the size, types of materials, and the integration of microscopic structures which control the modulation of the frequencies. These materials are then able to shield acoustic signals and attenuate the effects of anti-plane shear waves. By extrapolating these properties to larger scales it could be possible to create seismic wave filters.

According to research prior to this analysis, arrayed metamaterials can create filters or polarizers of either electromagnetic or elastic waves. Here a method is shown which can be applied to two-dimensional stop band and bandgap control with either photonic or sonic structures. Similar to photonic and electromagnetic metamaterial fabrication, a sonic metamaterial is embedded with localized sources of mass density ρ and the (elastic) bulk modulus β parameters, which are analogous to permittivity and permeability, respectively. The sonic (or phononic) metamaterials are sonic crystals, as in the previous section. These crystals have a solid lead core and a softer, more elastic silicone coating. The sonic crystals had built-in localized resonances due to the coated spheres which resulted in almost flat dispersion curves. Low-frequency bandgaps and localized wave interactions of the coated spheres were analyzed and presented in.

This method can be used to tune bandgaps inherent in the material and, also, create new low-frequency bandgaps. It is also applicable for designing low-frequency phononic crystal waveguides (radio frequency). Doubly periodic square array of SRRs are used to illustrate the methodology.

Phononic Crystal

Phononic crystals are synthetic materials that are formed by periodic variation of the acoustic properties of the material (i.e., elasticity and mass). One of the main properties of the phononic crystals is the possibility of having a phononic bandgap. A phononic crystal with phononic bandgap prevents phonons of selected ranges of frequencies from being transmitted through the material.

To obtain the frequency band structure of a phononic crystal, Bloch theory is applied on a single unit cell in the reciprocal lattice space (Brillouin zone). Several numerical methods are available for this problem, e.g., the planewave expansion method, the finite element method, and the finite difference method. A brief survey of numerical methods for calculating the frequency band structure is provided by Hussein (2009) In order to speed up the calculation of the frequency band structure, the Reduced Bloch Mode Expansion (RBME) method can be used. The RBME applies "on top" of any of the primary expansion numerical methods mentioned above. For large unit cell models, the RBME method can reduce the time for computing the band structure by up to two orders of magnitude. The basis of phononic crystals dates back to Isaac Newton who imagined that sound waves propagated through air in the same way that an elastic wave would propagate along a lattice of point masses connected by springs with an elastic force constant E. This force constant is identical to the modulus of the material. Of course with phononic crystals of materials with differing modulus the calculations are a little more complicated than this simple model. Based on Newton's observation we can conclude that a key factor for acoustic band-gap engineering is impedance mismatch between periodic elements comprising the crystal and the surrounding medium. When an advancing wave-front meets a material with very high impedance it will tend to increase its phase velocity through that medium. Likewise, when the advancing wave-front meets a low impedance medium it will slow down. We can exploit this concept with periodic (and handcrafted) arrangements of impedance mismatched elements to affect acoustic waves in the crystal – essentially band-gap engineering.

The position of the band-gap in frequency space for a phononic crystal is controlled by the size and arrangement of the elements comprising the crystal. The width of the band gap is generally related to the difference in the speed of sound (due to impedance differences) through the materials that comprise the composite.

Double-Negative Acoustic Metamaterial

The electromagnetic (isotropic) metamaterials have built-in resonance structures that exhibit effective negative permittivity and negative permeability for some frequency ranges. In contrast, it is difficult to build composite acoustic materials with built-in resonances such that the two effective response functions are negative within the capability or range of the transmission medium.

The mass density ρ and bulk modulus β are position dependent. Using the formulation of a plane wave the wave vector is:

$$\vec{k} = \frac{|n|\omega}{c}.$$

The angular frequency is represented by ω and *c* is the propagation speed of acoustic signal through the homogeneous medium. With constant density and bulk modulus as constituents of the medium, the refractive index is expressed as $n^2 = \rho / \beta$. In order to develop a propagating (plane) wave through the material, it is necessary for both ρ and β to be either positive or negative.

When the negative parameters are achieved, the mathematical result of the Poynting vector $\vec{s}$. is the opposite direction of the wave vector $\vec{k}$. This requires negativity in bulk modulus and density. Physically, it means that the medium displays an anomalous response at some frequencies such that it expands upon compression (negative bulk modulus) and moves to the left when being pushed to the right (negative density) at the same time. Natural materials do not have a negative density or a negative bulk modulus, but, negative values are mathematically possible, and can be demonstrated when dispersing soft rubber in a liquid. Even for composite materials, the effective bulk modulus and density should be normally bounded by the values of the constituents, i.e., the derivation of lower and upper bounds for the elastic moduli of the medium. Intrinsic is the expectation for positive bulk modulus and positive density. For example, dispersing spherical solid particles in a fluid results in the ratio governed by the specific gravity when interacting with the long acoustic wavelength (sound). Mathematically, it can be easily proven that β_{eff} and ρ_{eff} are definitely positive for natural materials. The exception occurs at low resonant frequencies.

As an example, acoustic double negativity is theoretically demonstrated with a composite of soft, silicone rubber spheres suspended in water. In soft rubber, sound travels much slower than through the water. The high velocity contrast of sound speeds between the rubber spheres and the water allows for the transmission of very low monopolar and dipolar frequencies. This is an analogue to analytical solution for the scattering of electromagnetic radiation, or electromagnetic plane wave scattering, by spherical particles - dielectric spheres. Hence, there is a narrow range of normalized frequency $0.035 < \omega a/(2\pi c) < 0.04$ where the bulk modulus and negative density are both negative. Here *a* is the lattice constant if the spheres are arranged in a face-centred cubic (fcc) lattice; ω is frequency and *c* is speed of the acoustic signal. The effective bulk modulus and density near the static limit are positive as predicted. The monopolar resonance creates a negative bulk modulus above the normalized frequency at about 0.035 while the dipolar resonance creates a negative density above the normalized frequency at about 0.04.

This behaviour is analogous to low-frequency resonances produced in SRRs (electromagnetic metamaterial). The wires and split rings create intrinsic electric dipolar and magnetic dipolar response. With this artificially constructed acoustic metamaterial of rubber spheres and water, only one structure (instead of two) creates the low-frequency resonances to achieve double negativity.

With monopolar resonance, the spheres expand, which produces a phase shift between the waves passing through rubber and water. This creates the negative response. The dipolar resonance creates a negative response such that the frequency of the centre of mass of the spheres is out of phase with the wave vector of the sound wave (acoustic signal). If these negative responses are large enough to compensate the background fluid, one can have both negative effective bulk modulus and negative effective density.

Both the mass density and the reciprocal of the bulk modulus are decreasing in magnitude fast enough so that the group velocity becomes negative (double negativity). This gives rise to the desired results of negative refraction. The double negativity is a consequence of resonance and the resulting negative refraction properties.

Metamaterial with Simultaneously Negative Bulk Modulus and Mass Density

In August 2007 a metamaterial was reported which simultaneously possesses a negative bulk modulus and mass density. This metamaterial is a zinc blende structure consisting of one fcc array of bubble-contained-water spheres (BWSs) and another relatively shifted fcc array of rubber-coated-gold spheres (RGSs) in special epoxy.

Negative bulk modulus is achieved through monopolar resonances of the BWS series. Negative mass density is achieved with dipolar resonances of the gold sphere series. Rather than rubber spheres in liquid, this is a solid based material. This is also as yet a realisation of simultaneously negative bulk modulus and mass density in a solid based material, which is an important distinction.

Double C Resonators

Double C resonator (DCR) is a ring cut in halves. In 2007, proposals have been made for arrays of DCRs and similar negative acoustic metamaterial. Although linear elasticity is mentioned, the problem is defined around shear waves directed at angles to the plane of the cylinders. The DCR was constructed similar to the SRRs in a multiple cell configuration. The DCR has been improved with stiffer material sheets. Each cell consists of a large rigid disk and two thin ligaments. The DCR cell is a tiny oscillator connected by springs. One spring of the oscillator connects to the *mass* and is anchored by the other spring. The LC resonator has specified capacitance and inductance. The limitations are expressed with appropriate mathematical equations. In addition to the intended limitations is that the speed of sound in the matrix is expressed as c = “ρ/μ with a matrix of density ρ and shear modulus μ. The resonant frequency is then expressed as “1/(LC).

A phononic bandgap occurs in association with the resonance of the split cylinder ring. There is a phononic band gap within a range of *normalized frequencies*. This is when the inclusion moves as a rigid body.

The DCR design produced a suitable band with negative slope in a range of frequencies. This band was obtained by hybridizing the modes of a DCR with the modes of thin stiff bars. Calculations have shown that at these frequencies:

- a beam of sound negatively refracts across a slab of such a medium,
- the phase vector in the medium possesses real and imaginary parts with opposite signs,
- the medium is well impedance-matched with the surrounding medium,
- a flat slab of the metamaterial can image a source across the slab like a Veselago lens,
- the image formed by the flat slab has considerable sub-wavelength image resolution, and
- a double corner of the metamaterial can act as an open resonator for sound.

Acoustic Metamaterial Superlens

In May 2009 Shu Zhang *et al.* presented the design and test results of an ultrasonic metamaterial lens for focusing 60 kHz (~2 cm wavelength) sound waves under water. The lens is made of sub-wavelength elements and is therefore potentially more compact than phononic lenses that operate in the same frequency range.

High-resolution acoustic imaging techniques are the essential tools for nondestructive testing and medical screening. However, the spatial resolution of the conventional acoustic imaging methods is restricted by the incident wavelength of ultrasound. This is due to the quickly fading evanescent fields which carry the sub-wavelength features of objects.

The lens consists of a network of fluid-filled cavities called Helmholtz resonators that oscillate at certain sonic frequencies. Similar to a network of inductors and capacitors in electromagnetic metamaterial, the arrangement of Helmholtz cavities designed by Zhang *et al.* have a negative dynamic modulus for ultrasound waves. Zhang *et al.* did focus a point source of 60.5 kHz sound to a spot size that is roughly the width of half a wavelength and their design may allow to push the spatial resolution even further. This result is in excellent agreement with the numerical simulation by transmission line model, which derived the effective mass density and compressibility. This metamaterial lens also displays variable focal length at different frequencies.

Bibliography

Altmann, J. & Gubrud, M.A.: *Discovering the Nanoscale*, IOS Press, Amsterdam, 2004.

Andrew, N.: *Foundations of Nanomechanics: from Solid-state Theory to Device Applications*, Berlin, New York, 2003.

Attebery, B.: *Nanoculture: Implications of the New Technoscience*, Intellect Books, Bristol, 2004.

Bainbridge, W.S. : *Societal Implications of Nanoscience and Nanotechnology*, Dordrecht, Kluwer, 2001.

Bast, R.C. Jr.: *Cancer Medicine*, Decker, Hamilton, 2000.

Berube, D.M.: *Discovering the Nanoscale*, IOS Press, Amsterdam, 2004.

Brosnan, J.: *The Primal Scream: A History of Science Fiction Film*, Orbit Books, London, 1991.

Carroll, J.S.: *Social Science Research Methods for Assessing Societal Implications of Nanotechnology*, Dordrecht, Kluwer, 2001.

Caviedes, C.: *El Nino in History: Storming Through the Ages*, University Press of Florida, Florida, 2001.

Credi, A.: *Molecular Devices and Machines: a Journey into the Nano World*, Wiley-VCH, Weinheim, 2003.

Crow, M. & Sarewitz, D.: *Societal Implications of Nanoscience and Nanotechnology*, Dordrecht: Kluwer, 2001.

Daniel Ratner: *Nanotechnology: a Gentle Introduction to the Next Big Idea*, Upper Saddle River, Prentice Hall, 2003.

Dobzhanshy, T.: *Genetics of the Evolutionary Process*, Columbia University Press, New York, 1970.

Drexler, K. Eric: *Engines of Creation: The Coming Era of Nanotechnology*, Anchor Books, New York, 1986.

Dyson, Tim: *India's Historical Demography: Studies in Famine, Disease, and Society*, Riverdale, Westwood, MA, 1989.

Eamonn Gearon: *The Sahara: A Cultural History*, Signal Books, UK, 2011.

Frank J. Owens: *Introduction to Nanotechnology*, Hoboken, J. Wiley, 2003.

Glimell, H.: *Dynamics of the Emerging Field of Nanoscience*, Dordrecht, Kluwer, 2001.

Glimell, H.: *Societal Implications of Nanoscience and Nanotechnology*, Kluwer, Dordrecht, 2001.

Haber, L. F.: *The Chemical Industry: 1900-1930, International Growth and Technological Change*, Oxford: Claredon Press, 1971.

Hessenbruch, A.: *Nanotechnology and the Negotiation of Novelty*, Amsterdam, IOS Press, 2004.

James Lewis: *Prospects in Nanotechnology: Toward Molecular Manufacturing,* John Wiley & Sons, New York, 1995.

Karn, Barbara: *Nanotechnology and the Environment: Applications and Implications,* American Chemical Society, Washington D.C., 2004.

Keblinski, P. : *Encyclopedia of Nanoscience and Nanotechnology*, American Scientific, Los Angeles, Calif, USA, 2004.

Liu W.K., Karpov E.G., Park H.S.: *Nano Mechanics and Materials,* Wiley, Weinheim, 2006.

Mayr, E.: *Change of Genetic Environment and Evolution*, Allen and Unwin, London, 1954.

Mody, C.C.M.: *How Probe Microscopists Became Nanotechnologists, Discovering the Nanoscale*, IOS Press, Amsterdam, 2004.

Norio Taniguchi: *Nanotechnology: Integrated Processing Systems for Ultra-precision and Ultra-fine Products,* Oxford University Press, New York, 1996.

Orth, R. N.: *Nanometer Scale Antibody Patterning for Directed cell Immobilization and Stimulation,* Cornell University, Ithaca, 2002.

Pressman, J.: *Nano Narrative: A Parable from Electronic Literature*, Intellect Books, Bristol, 2004.

Pressman, J.: *Nanoculture: Implications of the New Technoscience*, Intellect Books, Bristol, UK, 2004.

Ratner, Daniel: *Nanotechnology: a Gentle Introduction to the Next Big Idea*, Prentice Hall, Upper Saddle River, 2003.

Robison, W.L.: *Discovering the Nanoscale*, IOS Press, Amsterdam, 2004.

Sarewitz, D.: *Nanotechnology and Societal Transformation*, Dordrecht, Kluwer, 2001.

Stephen M. Goodnick: *Transport in Nanostructures,* Cambridge University Press, Cambridge, 1997.

Tenner, E.: *Nanotechnology and Unintended Consequences*, , Dordrecht, Kluwer, 2001.

Vasudev Bhatnagar: *Cell Science and Technology*, Campus Books, Delhi, 2009.

Vermaas, P.E.: *Discovering the Nanoscale*, IOS Press, Amsterdam, 2004.

Weil, V.: *Ethical Issues in Nanotechnology*, Dordrecht, Kluwer, 2001.

Index

K

L

M

N

O

P

Q

R

S

T

U

V

W

X

Y

Z

❑❑❑